INDUSTRIAL MICROBIOLOGY

INDUSTRIAL MICROBIOLOGY

Banwari Lal

CYBER TECH PUBLICATIONS
4264/3, Ansari Road, Daryaganj, New Delhi-110002 (India)
Ph.: 011-23244078. Fax: 011-23280028
E-Mail: cyberpublicationsdelhi@yahoo.com

INDUSTRIAL MICROBIOLOGY

Banwari Lal

First Edition 2009

Published by :

G.S. Rawat for **Cyber Tech Publications**
4264/3, Ansari Road, Daryaganj, New Delhi-110002 (India)
Ph.: 011-23244078. Fax: 011-23280028
E-Mail: cyberpublicationsdelhi@yahoo.com

Printed at : **Amit Enterprises**, Maujpur, Delhi, (9810878184)

Preface

Microbiology is an exceptionally broad discipline encompassing specialties as diverse as biochemistry, cell biology, genetics, taxonomy, pathogenic bacteriology, food and industrial microbiology, and ecology.

Microbiology is an exceptionally broad discipline encompassing specialties as diverse as biochemistry, cell biology, genetics, taxonomy, pathogenic bacteriology, food and industrial microbiology, and ecology. A microbiologist must be acquainted with many biological disciplines and with all major groups of microorganisms: viruses, bacteria, fungi, algae, and protozoa. The key is balance. Students new to the subject need an introduction to the whole before concentrating on those parts of greatest interest to them. This text provides a balanced introduction to all major areas of microbiology for a variety of students. Because of this balance, the book is suitable for courses with orientations ranging from basic microbiology to medical and applied microbiology. Students preparing for careers in medicine, dentistry, nursing, and allied health professions will find the text just as useful as those aiming for careers in research, teaching, and industry. The book is organized flexibly so that chapters and topics may be arranged in almost any order. Each chapter has been made as self-contained as possible to promote this flexibility. Some topics are essential to microbiology and have been given more extensive treatment.

— *Author*

CONTENTS

1

Introduction

Industrial microbiology or microbial biotechnology encompasses the use of microorganisms in the manufacture of food or industrial products. The use of microorganisms for the production of food, either human or animal, is often considered a branch of food microbiology. The microorganisms used in industrial processes may be natural isolates, laboratory selected mutants or genetically engineered organisms.

FOOD MICROBIOLOGY

Yogurt, cheese, chocolate, butter, pickles, sauerkraut, soya sauce, vitamins, amino acids, food thickeners (microbial polysaccharides), alcohol, sausages, and silage (animal food) are all produced by industrial microbiology processes. "Good" bacteria such as probiotics are becoming increasingly important in the food industry.

Biopolymers

A huge variety of biopolymers, such as polysaccharides, polyesters, and polyamides, are produced by micro-organisms. These products range from viscous solutions to plastics. The genetic manipulation of microorganisms has permitted the biotechnological production of biopolymers with tailored material properties suitable for high-value medical application such as tissue engineering and drug delivery.

Industrial microbiology can be used for the biosynthesis of xanthan, alginate, cellulose, cyanophycin, poly(gamma-glutamic acid), levan, hyaluronic acid, organic acids, oligosaccharides and polysaccharides, and polyhydroxyalkanoates.

Bioremediation

Microbial biodegradation of pollutants can be used to cleanup contaminated environments.

These bioremediation and biotransformation methods harness naturally occurring microbes to degrade, transform or accumulate a huge range of compounds including hydrocarbons (e.g. oil), polychlorinated biphenyls (PCBs), polyaromatic hydrocarbons (PAHs), pharmaceutical substances, radionuclides and metals.

Waste Biotreatment

Microorganisms are used to treat the vast quantities of wastes generated by modern societies. Biotreatment, the processing of wastes using living organisms, is an environmentally friendly, relatively simple and cost-effective alternative to physico-chemical clean-up options. Confined environments, such as bioreactors] can be employed in biotreatment processes.

Health-care and Medicine

Microorganisms are used to produce human or animal biologicals such as insulin, growth hormone, and antibodies. Diagnostic assays that use monoclonal antibody, DNA probe technology or real-time PCR are used as rapid tests for pathogenic organisms in the clinical laborarory.

Archaea

Examination of microbes living in unusual environments (e.g. high temperatures, salt, low pH or temperature, high radiation) an lead to discovery of microbes with new abilities that can be harnessed for industrial purposes.

Corynebacteria

Corynebacteria are a diverse group Gram-positive bacteria found in a range of different ecological niches such as soil, vegetables, sewage, skin, and cheese smear. *Corynebacterium glutamicum* is of immense industrial importance and is one of the biotechnologically most important bacterial species with an annual production of more than two million tons of amino acids, mainly L-glutamate and L-lysine. The genome sequence of *C. glutamicum* has been published.

BIOREMEDIATION

Bioremediation can be defined as any process that uses microorganisms, fungi, green plants or their enzymes to return the natural environment altered by contaminants to its original condition. Bioremediation may be employed to attack specific soil contaminants, such as degradation of chlorinated hydrocarbons by bacteria. An example of a more general approach is the cleanup of oil spills by the addition of nitrate and/or sulfate fertilisers to facilitate the decomposition of crude oil by indigenous or exogenous bacteria.

Overview and Applications

Naturally occurring bioremediation and phytoremediation have been used for centuries. For example, desalination of agricultural land by phytoextraction has a long tradition. Bioremediation technology using microorganisms was reportedly invented by George M. Robinson. He was the assistant county petroleum engineer for Santa Maria, California. During the 1960's, he spent his spare time experimenting with dirty jars and various mixes of microbes.

Bioremediation technologies can be generally classified as *in situ* or *ex situ*. *In situ* bioremediation involves treating the contaminated material at the site while *ex situ* involves the removal of the contaminated material to be treated elsewhere. Some examples of bioremediation technologies are bioventing, landfarming, bioreactor, composting, bioaugmentation, rhizofiltration, and biostimulation.

Not all contaminants, however, are easily treated by bioremediation using microorganisms. For example, heavy metals such as cadmium and lead are not readily absorbed or captured by organisms. The assimilation of metals such as mercury into the food chain may worsen matters. Phytoremediation is useful in these circumstances, because natural plants or transgenic plants are able to bioaccumulate these toxins in their above-ground parts, which are then harvested for removal. The heavy metals in the harvested biomass may be further concentrated by incineration or even recycled for industrial use ability of organisms to adapt to changing environmental conditions.

The elimination of a wide range of pollutants and wastes from the environment is an absolute require increasing our understanding of the relative importance of different pathways and regulatory networks to carbon flux in particular environments and for particular compounds and they will certainly accelerate the development of bioremediation technologies and biotransformation processes.

Genetic Engineering Approaches

The use of genetic engineering to create organisms specifically designed for bioremediation has great potential. The bacterium Deinococcus radiodurans (the most radioresistant organism known) has been modified to consume and digest toluene and ionic mercury from highly radioactive nuclear waste.

Advantages

There are a number of cost/efficiency advantages to bioremediation, which can be employed in areas that are inaccessible without excavation. For example, hydrocarbon spills (specifically, petrol spills) or certain chlorinated solvents may contaminate groundwater, and introducing the appropriate electron acceptor or electron donor amendment, as appropriate, may significantly reduce contaminant concentrations after a lag time allowing for acclimation. This is typically much less expensive than excavation followed by disposal elsewhere, incineration or other *ex situ* treatment strategies, and reduces or eliminates the need for "pump and treat", a common practice at sites where hydrocarbons have clean groundwater.

Monitoring Bioremediation

The process of bioremediation can be monitored indirectly by measuring the *Oxidation Reduction Potential* or redox in soil and groundwater, together with pH, temperature, oxygen content, electron acceptor/donor concentrations, and concentration of breakdown products (e.g. carbon dioxide). This table shows the (decreasing) biological breakdown rate as function of the redox potential.

This, by itself and at a single site, gives little information about the process of remediation.

1. it is necessary to sample enough points on and around the contaminated site to be able to determine contours of equal redox potential. Contouring is usually done using specialised software, e.g. using Kriging interpolation.
2. if all the measurements of redox potential show is that electron acceptors have been used up, it's in effect an indicator for total microbial activity. Chemical analysis is also required to determine when the levels of contaminants and their breakdown products have been reduced to below regulatory limits.

BIODEGRADATION

Biodegradation is the process by which organic substances are broken down by the enzymes produced by living organisms. The term is often used in relation to ecology, waste management and environmental remediation (bioremediation). Organic material can be degraded aerobically, with oxygen, or anaerobically, without oxygen. A term related to biodegradation is biomineralisation, in which organic matter is converted into minerals.

Biodegradable matter is generally organic material such as plant and animal matter and other substances originating from living organisms, or artificial materials that are similar enough to plant and animal matter to be put to use by microorganisms. Some microorganisms have the astonishing, naturally occurring, microbial catabolic diversity to degrade, transform or accumulate a huge range of compounds including hydrocarbons (e.g. oil), polychlorinated biphenyls (PCBs), polyaromatic hydrocarbons (PAHs), pharmaceutical substances, radionuclides and metals. Major methodological breakthroughs in microbial biodegradation have enabled detailed genomic, metagenomic, proteomic, bioinformatic and other high-throughput analyses of environmentally relevant microorganisms providing unprecedented insights into key biodegradative pathways and the ability of microorganisms to adapt to changing environmental conditions.

Anaerobic Biodegradation in Landfill

Biodegradable waste in landfill degrades in the absence of oxygen through the process of anaerobic digestion. The byproducts of this anaerobic biodegradation are biogas and lignin and cellulose fibres which cannot be broken down by anaerobes (anaerobic microbes)

Engineered landfills are designed with liners to prevent toxic leachate seeping into the surrounding soil and groundwater. Paper and other materials that normally degrade in a few years degrade more slowly over longer periods of time. Biogas contains methane which has approximately 21 times the global warming potential of carbon dioxide. In modern landfills this biogas can be collected and used for power generation.

Methods of Measuring Biodegradation

Biodegradation can be measured in a number of ways. The activity of aerobic microbes can be measured by the amount of oxygen they consume or the amount of carbon dioxide they produce. Biodegradation can be measured by anaerobic microbes and the amount of methane or alloy that they may be able to produce.

Measurement of Aerobic Decomposition

The DR4 test or 4-day dynamic respiration index test is a test to measure the biodegradability of a substance over

4 days. The substance is aerated by passing air through it. This definition is used to determine the method from those where aeration is by diffusion of air into and out of the test material which is referred to as the SRI or static respiration index test. Microbes are introduced to the test material while incubating it under aerobic conditions by aerating the mixture in a vessel through which air is blown. The microbes degrade the material producing CO_2 as the product of biodegradation. This CO_2 production can be monitored as a measure of the biodegradability of the test material and converted into oxygen consumption units.

Measurement of Anaerobic Decomposition

BMP100 test, 100 day biogenic methane potential test, is a test method that determines the potential biodegradability of biodegradable wastes under anaerobic conditions by measuring the production of biogas. The test has not been peer-reviewed by the international community and no known official publication exists for it. Other published tests that accomplish this in shorter time are the GB21 protocol (DIN 38414).

Under anaerobic methanogenic conditions the decomposition of organic carbon proceeds by producing biogas (containing methane and carbon dioxide)from the organic carbon. The amount of biogas production therefore measures directly the carbon which is mineralised. The test is set up in a small vessel containing the test substrate, a mineral aqueous medium and an inoculum of methanogenic bacteria taken from an active anaerobic digester. The test is monitored by collecting and measuring the biogas produced. The test is incubated for an extended period until gas production ceases which may be up to 100 days or more; for all practical purposes most organic materials reach the majority of decomposition in less than 45 days. By being run so long, however, the BMP100 test therefore measures the complete degradation of the waste.

Plastics

Biodegradable plastics made with plastarch material (PSM), and polylactide (PLA) will compost in an industrial compost facility. There are other plastic materials that claim biodegradability, but are more often (and possibly more accurately) described as 'degradable' or oxi-degradable; It is claimed that this process causes more rapid breakdown of the plastic materials into CO_2 and H_2O.

Indicative Lengths of Degradation

The following table should be read with the above comments in mind, and care should be taken before accepting claims of biodegradability in view of the (dubious) claims being made. This is how long it takes for some commonly used products to biodegrade:

- Banana peel, 2 – 10 days
- Cotton rags, 1 – 5 months
- Sugarcane Pulp Products, 30 - 60 days
- Paper, 2 – 5 months
- Rope, 3 – 14 months
- Orange peels, 6 months
- Wool socks, 1 – 5 years
- Cigarette filters, 1 – 12 years
- Tetrapaks (plastic composite milk cartons), 5 years
- Plastic bags, 10 – 20 years
- Leather shoes, 25 – 40 years

- Nylon fabric, 30 – 40 years
- Plastic six-pack holder rings, 450 years
- Diapers and sanitary napkins 500 – 800 years
- Tin cans 50 - 100 years
- Aluminum cans 80 - 100 years
- Plastic Bottles non-biodegradeable
- Styrofoam cup, non-biodegradeable

Some Facts About Regular Plastic Bags:

1. Many countries throughout Europe have imposed an "eco-tax" on regular plastic bags. Biodegradable plastic bags are exempt from such taxes.
2. Regular plastic bags take up to 1,000 years to degrade, and even then don't ever really dissappear.
3. It has been estimated that it costs governments and business over $4 Million per year to clean up littering plastic shopping bags.
4. An estimated 300 billion plastic bags, or 1.8 million tons of plastic bags are disposed of in landfill sites throughout the US alone, each year.
5. Plastics are made from non-renewable natural resources such as oil, gas, and coal. The manufacturing processes of making plastic bags pollute the environment with harmful greenhouse gases.
6. There are approximately 46,000 pieces of plastic floating in each square mile of our oceans.
7. Just 14 grocery bags contain enough embodied petroleum energy to drive a car 1 mile.
8. Each year, Americans use 117 billion new plastic bags. If these were all tied together, they would stretch around the world 725 times!

9. Americans dump 68,000 plastic bags into landfill every minute. That's 4,080,000 bags every hour!
10. It is estimated that plastic kills 1 million sea birds, 100,000 sea mammals and countless fish each year.
11. Turtles, dolphins and killer whales can choke or starve by confusing plastic bags for jellyfish. See the devastating effects plastic bags have on the environment.

How Long Does It Take For Other Products To Biodegrade?

Here's how long it takes for some commonly used products to biodegrade, after they are disposed of:

Cotton rags—1-5 months

Paper—2-5 months

Rope—3-14 months

Orange peels—6 months

Wool socks—1-5 years

Cigarette butts—1-12 years

Plastic coated paper milk cartons—5 years

Leather shoes—25-40 years

Nylon fabric—30-40 years

Tin cans—50-100 years

Aluminium cans—80-100 years

Plastic 6-pack holder rings—450 years

Glass bottles—1 million years

Plastic bottles—Forever

Impact on Wildlife

The real impact of plastic bag litter affects wildlife both in the marine environment and in rural areas. Tens of thousands of whales, birds, seals and turtles are killed every year from plastic bag litter in the marine environment as they often mistake plastic bags for food such as jellyfish. Plastic bags, once ingested, cannot be digested or passed by an animal so it stays in their stomachs.

Plastic in an animal's digestive tract can prevent food digestion and can lead to a very slow and painful death. In the case of biodegradable plastic bags, the bags would disolve in water after a short time, while those bags that would be swallowed by marine animals, the animals' digestive fluids would dissolve the biodegradable plant additive, therefore breaking the bag up into little pieces, which the animal could later pass naturally.

As plastic bags can take up to 1000 years to break down, once an animal dies and decays after ingesting plastic, the plastic is then freed back into the marine environment to carry on killing other wildlife! Turtles have also been rescued with plastic bags lodged in their throat – and part of the bag hanging out of their mouth. The turtle pictured above was found and rescued by a senior staff member of Melbourne (Australia) Zoo while they were out doing fieldwork.

ANAEROBIC DIGESTION

Anaerobic digestion is a process in which micro-organisms break down biodegradable material in the absence of oxygen. The process is widely used to treat wastewater sludges and organic wastes because it provides volume and mass reduction of the input material. As part of an integrated waste management system, anaerobic digestion reduces the emission of landfill gas into the atmosphere. Anaerobic digestion is a renewable energy source because the process produces a methane and carbon dioxide rich biogas suitable for energy production helping replace fossil fuels. Also, the nutrient-rich solids left after digestion can be used as fertiliser.

The digestion process begins with bacterial hydrolysis of the input materials in order to break down insoluble organic polymers such as carbohydrates and make them available for other bacteria. Acidogenic bacteria then convert the sugars and amino acids into carbon dioxide, hydrogen, ammonia, and organic acids.

Acetogenic bacteria then convert these resulting organic acids into acetic acid, along with additional ammonia, hydrogen, and carbon dioxide. Methanogenic bacteria finally are able to convert these products to methane and carbon dioxide.

The technical expertise required to maintain anaerobic digesters coupled with high capital costs and lower process efficiencies have so far limited the level of its industrial application as a waste treatment technology.

Anaerobic digestion facilities have, however, been recognised by the United Nations Development Programme as one of the most useful decentralised sources of energy supply, as they are less capital intensive than large powerplants.

Careful control of the digestion temperature, pH, and loading rates is crucial to obtaining efficient breakdown of the material, and disturbances to a digest can lead to process failure. Ensuring that the quality of input materials to the digesters is maintained and that the process effectively monitored is essential for ensuring that a digester's performance is reliable.

History

Scientific interest in the gases produced by the natural decomposition of organic matter, was first reported in the sixteenth century by Robert Boyle and Stephen Hale, who noted that inflammable gas was released by disturbing the sediment of streams and lakes. In 1808, Sir Humphry Davy determined that methane was present in the gases produced by cattle manure.

The first anaerobic digester was built by a leper colony in Bombay, India in 1859. In 1895 the technology was developed in Exeter, England, where a septic tank was used to generate gas for street lighting. Also in England, in 1904, the first dual purpose tank for both sedimentation and sludge treatment was installed in Hampton. In 1907, in Germany, a patent was issued for the Imhoff tank, an early form of digester.

Through scientific research anaerobic digestion gained academic recognition in the 1930s. This research led to the discovery of anaerobic bacteria, the microorganisms that facilitate the process. Further research was carried out to investigate the conditions under which methanogenic bacteria were able to grow and reproduce. This work was developed during World War II where in both Germany and France there was an increase in the application of anaerobic digestion for the treatment of manure.

Applications

Anaerobic digestion is particularly suited to wet organic material and is commonly used for effluent and sewage treatment. Anaerobic digestion is a simple process that can greatly reduce the amount of organic matter which might otherwise be destined to be landfilled or burnt in an incinerator.

Almost any organic material can be processed with anaerobic digestion. This includes biodegradable waste materials such as waste paper, grass clippings, leftover food, sewage and animal waste. The exception to this is woody wastes that are largely unaffected by digestion as anaerobes are unable to degrade lignin. Anaerobic digesters can also be fed with specially grown energy crops such as silage for dedicated biogas production. In Germany and continental Europe these facilities are referred to as *biogas plants*. A *co-digestion* or *co-fermentation* plant is typically an agricultural anaerobic digester that accepts two or more input materials for simultaneous digestion.

In developing countries simple home and farm-based anaerobic digestion systems offer the potential for cheap, low-cost energy for cooking and lighting. Anaerobic digestion facilities have been recognised by the United Nations Development Programme as one of the most useful decentralised sources of energy supply. From 1975, China and India have both had large government-backed schemes for adaptation of small biogas plants for use in the household for cooking and lighting. Presently, projects for anaerobic digestion in the developing world can gain financial support through the United Nations Clean Development Mechanism if they are able to show the provide reduced carbon emissions.

Pressure from environmentally-related legislation on solid waste disposal methods in developed countries has increased the application of anaerobic digestion as a process for reducing waste volumes and generating useful by-products. Anaerobic digestion may either be used to process the source separated fraction of municipal waste, or alternatively combined with mechanical sorting systems, to process residual mixed municipal waste. These facilities are called mechanical biological treatment plants. Utilising anaerobic digestion technologies can help to reduce the emission of greenhouse gases in a number of key ways:

- Replacement of fossil fuels
- Reducing methane emission from landfills
- Displacing industrially-produced chemical fertilisers
- Reducing vehicle movements
- Reducing electrical grid transportation losses

Methane and power produced in anaerobic digestion facilities can be utilised to replace energy derived from fossil fuels, and hence reduce emissions of greenhouse gases. This is due to the fact that the carbon in biodegradable material is part of a carbon cycle. The carbon released into the atmosphere from the combustion of biogas has been removed by plants in order for them to grow in the recent past. This can have occurred within the last decade, but more typically within the last growing season. If the plants are re-grown, taking the carbon out of the atmosphere once more, the system will be carbon neutral. This contrasts to carbon in fossil fuels that has been sequestered in the earth for many millions of years, the combustion of which increases the overall levels of carbon dioxide in the atmosphere.

If the putrescible waste processed in anaerobic digesters was disposed of in a landfill, it would break down naturally and often anaerobically. In this case the gas will eventually escape into the atmosphere. As methane is about twenty times more potent as a greenhouse gas as carbon dioxide this has significant negative environmental effects.

Digestate liquor can be used as a fertiliser supplying vital nutrients to soils. The solid, fibrous component of digestate can be used as a soil conditioner. The liquor can be used as a substitute for chemical fertilisers which require large amounts of energy to produce. The use of manufactured fertilisers is therefore more carbon intensive than the use of anaerobic digestate fertiliser. This solid digestate can be used to boost the organic content of soils. There are some countries, such as in Spain where there are many organically depleted soils, and here the markets for the digestate can be just as important as the biogas.

In countries that collect household waste, the utilisation of local anaerobic digestion facilities can help to reduce the amount of waste that requires transportation to centralised landfill sites or incineration facilities. This reduced burden on transportation has an will reduce carbon emissions from the collection vehicles. If localised anaerobic digestion facilities are embedded within an electrical distribution network, they can help reduce the electrical losses that is associated with transporting electricity over a national grid.

The Process

There are a number of bacteria that are involved in the process of anaerobic digestion including acetic acid-forming bacteria (acetogens) and methane-forming bacteria (methanogens). These bacteria feed upon the initial feedstock, which undergoes a number of different processes converting it to intermediate molecules including sugars, hydrogen & acetic acid before finally being converted to biogas.

Different species of bacteria are able to survive at different temperature ranges. Ones living optimally at temperatures between 35-40°C are called mesophiles or mesophilic bacteria. Some of the bacteria can survive at the hotter and more hostile conditions of 55-60°C, these are called thermophiles or thermophilic bacteria. Methanogens come from the primitive group of archaea. This family includes species that can grow in the hostile conditions of hydrothermal vents. These species are more resistant to heat and can therefore operate at thermophilic temperatures, a property that is unique to bacterial families.

As with aerobic systems the bacteria in anaerobic systems the growing and reproducing microorganisms within them require a source of elemental oxygen to survive.

In an anaerobic system there is an absence of gaseous oxygen. In an anaerobic digester, gaseous oxygen is prevented from entering the system through physical containment in sealed tanks. Anaerobes access oxygen from sources other than the surrounding air.

The oxygen source for these microorganisms can be the organic material itself or alternatively may be supplied by inorganic oxides from within the input material. When the oxygen source in an anaerobic system is derived from the organic material itself, then the 'intermediate' end products are primarily alcohols, aldehydes, and organic acids plus carbon dioxide.

In the presence of specialised methanogens, the intermediates are converted to the 'final' end products of methane, carbon dioxide with trace levels of hydrogen sulfide.[31][32] In an anaerobic system the majority of the chemical energy contained within the starting material is released by methanogenic bacteria as methane.

Populations of anaerobic bacteria typically take a significant period of time to establish themselves to be fully effective. It is therefore common practice to introduce anaerobic microorganisms from materials with existing populations. This process is called 'seeding' the digesters and typically takes place with the addition of sewage sludge or cattle slurry.

Stages

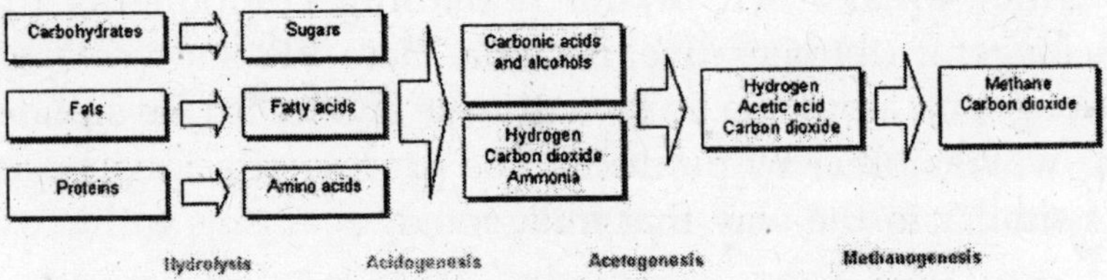

The key process stages of anaerobic digestion

There are four key biological and chemical stages of anaerobic digestion:

1. Hydrolysis
2. Acidogenesis
3. Acetogenesis
4. Methanogenesis

In most cases biomass is made up of large organic polymers. In order for the bacteria in anaerobic digesters to access the energy potential of the material, these chains must first be broken down into their smaller constituent parts. These constituent parts or monomers such as sugars are readily available by other bacteria. The process of breaking these chains and dissolving the smaller molecules into solution is called hydrolysis. Therefore hydrolysis of these high molecular weight polymeric components is the necessary first step in anaerobic digestion. Through hydrolysis the complex organic molecules are broken down into simple sugars, amino acids, and fatty acids.

Acetate and hydrogen produced in the first stages can be used directly by methanogens. Other molecules such as volatile fatty acids (VFA's) with a chain length that is greater than acetate must first be catabolised into compounds that can be directly utilised by methanogens.

The biological process of acidogenesis is where there is further breakdown of the remaining components by acidogenic (fermentative) bacteria. Here VFAs are created along with ammonia, carbon dioxide and hydrogen sulfide as well as other by-products. The process of acidogenesis is similar to the way that milk sours.

The third stage anaerobic digestion is acetogenesis. Here simple molecules created through the acidogenesis phase are further digested by acetogens to produce largely acetic acid as well as carbon dioxide and hydrogen.

The terminal stage of anaerobic digestion is the biological process of methanogenesis. Here methanogens utilise the intermediate products of the preceding stages and convert them into methane, carbon dioxide and water. It is these components that makes up the majority of the biogas emitted from the system. Methanogenesis is sensitive to both high and low pHs and occurs between pH 6.5 and pH 8. The remaining, non-digestable material which the microbes cannot feed upon, along with any dead bacterial remains constitutes the digestate.

A simplified generic chemical equation for the overall processes outlined above is as follows:

$$C_6H_{12}O_6 \rightarrow 3CO_2 + 3CH_4$$

Feedstock

The most important initial issue when considering the application of anaerobic digestion systems is the feedstock to the process. Digesters typically can accept any biodegradable material, however if biogas production is the aim, the level of putrescibility is the key factor in its successful application. The more putrescible the material the higher the gas yields possible from the system.

Substrate composition is a major factor in determining the methane yield and methane production rates from the digestion of biomass. Techniques are available to determine the compositional characteristics of the feedstock, whilst parameters such as solids, elemental and organic analyses are important for digester design and operation.

Anaerobes can breakdown material to varying degrees of success from readily in the case of short chain hydrocarbons such as sugars, to over longer periods of time in the case of cellulose and hemicellulose. Anaerobic microorganisms are unable to break down long chain woody molecules such as lignin.

Anaerobic digesters were originally designed for operation using sewage sludge and manures. Sewage and manure are not, however, the material with the most potential for anaerobic digestion as the biodegradable material has already had the energy content taken out by the animal that produced it.

A second consideration related to the feedstock wiii be moisture content. The wetter the material the more suitable it will be to handling with standard pumps instead of energy intensive concrete pumps and physical means of movement. Also the wetter the material, the more volume and area it takes up relative to the levels of gas that are produced.

The moisture content of the target feedstock will also affect what type of system is applied to its treatment. In order to use a high solids anaerobic digester for dilute feedstocks, bulking agents such as compost should be applied to increase the solid content of the input material. Another key consideration is the carbon:nitrogen ratio of the input material. This ratio is the balance of food a microbe requires in order to grown. The optimal C:N ratio for the 'food' a microbe is 20–30.

The level of contamination of the feedstock material is a key consideration. If the feedstock to the digesters has significant levels of physical contaminants such as plastic, glass or metals then pre-processing will be required in order for the material to be used. If it is not removed then the digesters can be blocked and will not function efficiently. It is with this logic in mind that mechanical biological treatment plants are designed. The higher the level of pre-treatment a feedstock requires, the more processing machinery will be required and hence the project will have higher capital costs.

After sorting or screening to remove any physical contaminants, such as metals and plastics, from the feedstock the material is often shredded, minced and mechanically or hydraulically pulped to increase the surface area available to microbes in the digesters and hence increase the speed of digestion. The feedstock material is then fed into the airtight digester where anaerobic treatment takes place.

Configuration

Farm-based maize silage digester located near Neumünster in Germany, 2007. Green inflatable biogas holder is shown on top of the digester. Anaerobic digesters can be designed and engineered to operate using a number of different process configurations:

- Batch or continuous
- Temperature: Mesophilic or thermophilic
- Solids content: High solids or low solids
- Complexity: Single stage or multistage

Batch or Continuous

A batch system is the simplest form of digestion. Biomass is added to the reactor at the start of the process in a batch and is sealed for the duration of the process. Batch reactors suffer from odour issues that can be a severe problem when they are emptied. Typically biogas production will be formed with a normal distribution pattern over time.

The operator can use this fact to determine when they believe the process of digestion of the organic matter has completed. As the batch digestion is simple and requires less equipment and lower levels of design work it is typically a cheaper form of digestion.

In continuous digestion processes organic matter is constantly or added in stages to the reactor. Here the end products are constantly or periodically removed, resulting in constant production of biogas. Examples of this form of anaerobic digestion include, continuous stirred-tank reactors (CSTRs), Upflow anaerobic sludge blanket (UASB), Expanded granular sludge bed (EGSB) and Internal circulation reactors (IC).

Temperature

There are two conventional operational temperature levels for anaerobic digesters, which are determined by the species of methanogens in the digesters:

- *Mesophilic* which takes place optimally around 37°-41°C or at ambient temperatures between 20°-45°C where mesophiles are the primary microorganism present
- *Thermophilic* which takes place optimally around 50°-52° at elevated temperatures up to 70°C where thermophiles are the primary microorganisms present

There are a greater number of species of mesophiles than thermophiles. These bacteria are also more tolerant to changes environmental conditions than thermophiles.

Mesophilic systems are therefore considered to be more stable than thermophilic digestion systems.

As mentioned above, thermophilic digestion systems are considered to be less stable, however the increased temperatures facilitate faster reaction rates and hence faster gas yields. Operation at higher temperatures facilitates greater sterilisation of the end digestate. In countries where legislation, such as the Animal By-Products Regulations in the European Union, requires end products to meet certain levels of reduction in the amount of bacteria in the output material, this may be a benefit.

A drawback of operating at thermophilic temperatures is that more heat energy input is required to achieve the correct operational temperatures. This increase in energy is not be outweighed by the increase in the outputs of biogas from the systems. It is therefore important to consider an energy balance for these systems.

Solids

Typically there are two different operational parameters associated with the solids content of the feedstock to the digesters:

- High-solids
- Low-solids

Digesters can either be designed to operate in a high solids content, with a total suspended solids (TSS) concentration greater than ~20%, or a low solids concentration less than ~15%.

High-solids digesters process a thick slurry that requires more energy input to move and process the feedstock. The thickness of the material may also lead to associated problems with abrasion. High-solids digesters will typically have a lower land requirement due to the lower volumes associated with the moisture.

Low-solids digesters can transport material through the system using standard pumps that require significantly lower energy input. Low-solids digesters require a larger amount of land than high-solids due to the increase volumes associated with the increased liquid: feedstock ratio of the digesters. There are benefits associated with operation in a liquid environment as it enables more thorough circulation of materials and contact between the bacteria and their food. This enables the bacteria to more readily access the substances they are feeding off and increases the speed of gas yields.

Number of Stages

Two-stage, low-solids, UASB digestion component of a mechanical biological treatment system near Tel Aviv, process water is seen in balance tank and sequencing batch reactor, 2005

Digestion systems can be configured with different levels of complexity:

- One-stage or single-stage
- Two-stage or multistage

A single-stage digestion system is one in which all of the biological reactions occur within a single sealed reactor or holding tank. Utilising a single stage reduces construction costs, however facilitates less control of the reactions occurring within the system. Acidogenic bacteria, through the production of acids, reduce the pH of the tank. Methanogenic bacteria, as outlined earlier, operate in a strictly defined pH range. Therefore the biological reactions of the different species in a single stage reactor can be in direct competition with each other. Another one-stage reaction system is an anaerobic lagoon. These lagoons are pond-like earthen basins used for the treatment and long-term storage of manures. Here the anaerobic reactions are contained within the natural anaerobic sludge contained in the pool.

In a two-stage or multi-stage digestion system different digestion vessels are optimised to bring maximum control over the bacterial communities living within the digesters. Acidogenic bacteria produce organic acids and more quickly grow and reproduce than methanogenic bacteria. Methanogenic bacteria require stable pH and temperature in order to optimise their performance.

Typically hydrolysis, acetogenesis and acidogenesis occur within the first reaction vessel. The organic material is then heated to the required operational temperature (either mesophilic or thermophilic) prior to being pumped into a methanogenic reactor.

The initial hydrolysis or acidogenesis tanks prior to the methanogenic reactor can provide a buffer to the rate at which feedstock is added. Some European countries require a degree of elevated heat treatment in order to kill harmful bacteria in the input waste. In this instance their may be a pasteurisation or sterilisation stage prior to digestion or between the two digestion tanks. It should be noted that it is not possible to completely isolate the different reaction phases and often there is some biogas that is produced in the hydrolysis or acidogenesis tanks.

Residence

The residence time in a digester varies with the amount and type of feed material, the configuration of the digestion system and whether it be one-stage or two-stage.

In the case of single-stage thermophilic digestion residence times may be in the region of 14 days, which comparatively to mesophilic digestion is relatively fast. The plug-flow nature of some of these systems will mean that the full degradation of the material may not have been realised in this timescale. In this event digestate exiting the system will be darker in colour and will typically have more odour.

In two-stage mesophilic digestion, residence time may vary between 15 and 40 days.

In the case of mesophilic UASB digestion hydraulic residence times can be (1hour-1day) and solid retention times can be up to 90 days. In this manner the UASB system is able to separate solid an hydraulic retention times with the utilisation of a sludge blanket.

Continuous digesters have mechanical or hydraulic devices, depending on the level of solids in the material, to mix the contents enabling the bacteria and the food to be in contact. They also allow excess material to be continuously extracted to maintain a reasonably constant volume within the digestion tanks.

Products

There are three principal products of anaerobic digestion: biogas, digestate and water.

Biogas

Biogas is the ultimate waste product of the bacteria feeding off the input biodegradable feedstock, and is mostly methane and carbon dioxide, with a small amount hydrogen and trace hydrogen sulfide. Most of the biogas is produced during the middle of the digestion, after the bacterial population has grown, and tapers off as the putrescible material is exhausted. The gas is normally stored on top of the digester in an inflatable gas bubble or extracted and stored next to the facility in a gas holder.

Typical composition of biogas

Matter	%
Methane, CH_4	50-75
Carbon dioxide, CO_2	25-50
Nitrogen, N_2	0-10
Hydrogen, H_2	0-1
Hydrogen sulfide, H_2S	0-3
Oxygen, O_2	0-2

The methane in biogas can be burned to produce both heat and electricity, usually with a reciprocating engine or microturbine often in a cogeneration arrangement where the electricity and waste heat generated are used to warm the digesters or to heat buildings. Excess electricity can be sold to suppliers or put into the local grid. Electricity produced by anaerobic digesters is considered to be renewable energy and may attract subsidies. Biogas does not contribute to increasing atmospheric carbon dioxide concentrations because the gas is not released directly into the atmosphere and the carbon dioxide comes from an organic source with a short carbon cycle.

Biogas may require treatment or 'scrubbing' to refine it for use as a fuel. Hydrogen sulfide is a toxic product formed from sulfates in the feedstock and is released as a trace component of the biogas. National environmental enforcement agencies such as the US EPA or the English and Welsh Environment Agency put strict limits on the levels of gases containing hydrogen sulfide, and if the levels of hydrogen sulfide in the gas are high, gas scrubbing and cleaning equipment (such as amine gas treating) will be needed to process the biogas to within regionally accepted levels. An alternative method to this is by the addition of ferric chloride $FeCl_3$ to the digestion tanks in order to inhibit hydrogen sulfide production.

Volatile siloxanes can also contaminate the biogas; such compounds are frequently found in household waste and wastewater. In digestion facilities accepting these materials as a component of the feedstock, low molecular weight siloxanes volatilise into biogas. When this gas is combusted in a gas engine, turbine or boiler, siloxanes are converted into silicon dioxide ($Si0_2$) which deposits internallv in the machine, increasing wear and tear.

In countries such as Switzerland, Germany and Sweden the methane in the biogas may be concentrated in order for it to be used as a vehicle transportation fuel or alternatively input directly into the gas mains. In countries where the driver for the utilisation of anaerobic digestion are renewable electricity subsidies, this route of treatment is less likely as energy is required in this processing stage and reduces the over all levels available to sell.

Digestate

Digestate is the solid remnants of the original input material to the digesters that the microbes cannot use. It also consists of the mineralised remains of the dead bacteria from within the digesters. Digestate can come in three forms; fibrous, liquor or a sludge-based combination of the two fractions. In two-stage systems the different forms of digestate come from different digestion tanks. In single stage digestion systems the two fractions will be combined and if desired separated by further processing.

The second by-product (acidogenic digestate) is a stable organic material comprised largely of lignin and cellulose, but also of a variety of mineral components in a matrix of dead bacterial cells; some plastic may be present. The material resembles domestic compost and can be used as compost or to make low grade building products such as fibreboard.

The third by-product is a liquid (methanogenic digestate) that is rich in nutrients and can be used as a fertiliser dependent on the quality of the material being digested.[83] Levels of potentially toxic elements (PTEs) should be chemically assessed.

This will be dependent upon the quality of the original feedstock. In the case of most clean and source-separated biodegradable waste streams the levels of PTEs will be low. In the case of wastes originating from industry the levels of PTEs may be higher and will need to be taken into consideration when determining a suitable end use for the material.

Digestate typically contains elements such as lignin that cannot be broken down by the anaerobic microorganisms. Also the digestate may contain ammonia that is phytotoxic and will hamper the growth of plants if it is used as a soil improving material.

For these two reasons a maturation or composting stage may be employed after digestion. Lignin and other materials are available for degradation by aerobic microorganisms such as fungi helping reduce the overall volume of the material for transport. During this maturation the ammonia will be broken down into nitrates, improving the fertility of the material and making it more suitable as a soil improver. Large composting stages are typically used by dry anaerobic digestion technologies.

Wastewater

The final output from anaerobic digestion systems is water. This water originates both from the moisture content of the original waste that was treated but also includes water produced during the microbial reactions in the digestion systems. This water may be released from the dewatering of the digestate or may be implicitly separate from the digestate.

The wastewater exiting the anaerobic digestion facility will typically have elevated levels of biochemical oxygen demand (BOD) and chemical oxygen demand (COD), these are measures of the reactivity of the effluent and show an ability to pollute. Some of this material is termed 'hard COD' meaning it cannot be accessed by the anaerobic bacteria for conversion into biogas. If this effluent was put directly into watercourses it would negatively affect them by causing eutrophication. As such further treatment of the wastewater is often required. This treatment will typically be an oxidation stage where air is passed through the water in a sequencing batch reactors or reverse osmosis unit.

2

Bioconversion of Biomass to Mixed Alcohol Fuels

The bioconversion of biomass to mixed alcohol fuels can be accomplished using the MixAlco process. Through bioconversion of biomass to a mixed alcohol fuel, more energy from the biomass will end up as liquid fuels than in converting biomass to ethanol by yeast fermentation.

The process involves a biological/chemical method for converting any biodegradable material (e.g., urban wastes, such as municipal solid waste, biodegradable waste, and sewage sludge, agricultural residues such as corn stover, sugarcane bagasse, cotton gin trash, manure) into useful chemicals, such as carboxylic acids (e.g., acetic, propionic, butyric acid), ketones (e.g., acetone, methyl ethyl ketone, diethyl ketone) and biofuels, such as a mixture of primary alcohols (e.g., ethanol, propanol, butanol) and/or a mixture of secondary alcohols (e.g., isopropanol, 2-butanol, 3-pentanol). Because of the many products that can be economically produced, this process is a true biorefinery .

The process uses a mixed culture of naturally occurring microorganisms found in natural habitats such as the rumen of cattle, termite guts, and marine and terrestrial swamps to anaerobically digest biomass into a mixture of carboxylic acids produced during the acidogenic and acetogenic stages of anaerobic digestion, however with the inhibition of the methanogenic final stage. The more popular methods for production of ethanol and cellulosic ethanol use enzymes that must be isolated first to be added to the biomass and thus convert the starch or cellulose into simple sugars, followed then by yeast fermentation into ethanol. This process does not need the addition of such enzymes as these microorganisms make their own .

As the microoganisms anaerobically digest the biomass and convert it into a mixture of carboxylic acids, the pH must be controlled. This is done by the addition of a buffering agent (e.g., ammonium bicarbonate, calcium carbonate), thus yielding a mixture of carboxylate salts. Methanogenesis, which, as mentioned, is the natural final stage of anaerobic digestion, is inhibited by the presence of the ammonium ions or by the addition of an inhibitor (e.g., iodoform). The resulting fermentation broth contains the produced carboxylate salts that must be dewatered. This is achieved efficiently by vapor-compression evaporation. Further chemical refining of the dewatered fermentation broth may then take place depending on the final chemical or biofuel product desired.

The condensed distilled water from the vapor-compression evaporation system is recycled back to the fermentation. On the other hand, if raw sewage or other waste water with high BOD in need of treatment is used as the water for the fermentation, the condensed distilled water from the evaporation can be recycled back to the city or to the original source of the high-BOD waste water. Thus, this process can also serve as a water treatment facility, while producing valuable chemicals or biofuels.

Because the system uses a mixed culture of microorganisms, besides not needing any enzyme addition, the fermentation requires no sterility or aseptic conditions, making this front step in the process more economical than in more popular methods for the production of cellulosic ethanol. These savings in the front end of the process, where volumes are large, allows flexibility for further chemical transformations after dewatering, where volumes are small.

Carboxylic Acids

Carboxylic acids can be regenerated from the carboxylate salts using a process known as "acid springing". This process makes use of a high-molecular-weight tertiary amine (e.g., trioctylamine), which is switched with the cation (e.g., ammonium or calcium). The resulting amine carboxylate can then be thermally decomposed into the amine itself, which is recycled, and the corresponding carboxylic acid. In this way, theoretically, no chemicals are consumed or wastes produced during this step.

Ketones

There are two methods for making ketones. The first one consists on thermally converting calcium carboxylate salts into the corresponding ketones. This was a common method for making acetone from calcium acetate during World War I. The other method for making ketones consists on converting the vaporized carboxylic acids on a catalytic bed of zirconium oxide .

Alcohols

Primary Alcohols

The undigested residue from the fermentation may bè used in gasification to make hydrogen (H_2). This H_2 can then be used to hydrogenolyze the esters over a catalyst (e.g., copper chromite), which are produced by esterifying either the ammonium carboxylate salts (e.g., ammonium acetate, propionate, butyrate) or the carboxylic acids (e.g., acetic, propionic, butyric acid) with a high-molecular-weight alcohol (e.g., hexanol, heptanol). From the hydrogenolysis, the final products are the high-molecular-weight alcohol, which is recycled back to the esterification, and the corresponding primary alcohols (e.g., ethanol, propanol, butanol).

Secondary Alcohols

The secondary alcohols (e.g., isopropanol, 2-butanol, 3-pentanol) are obtained by hydrogenating over a catalyst (e.g., Raney nickel) the corresponding ketones (e.g., acetone, methyl ethyl ketone, diethyl ketone).

Acetic Acid versus Ethanol

Cellulosic-ethanol -manufacturing plants are bound to be net exporters of electricity because a large portion of the lignocellulosic biomass, namely lignin, remains undigested and it must be burned, thus producing electricity for the plant and excess electricity for the grid. As the market grows and this technology becomes more widespread, coupling the liquid fuel and the electricity markets will become more and more difficult.

Acetic acid, unlike ethanol, is biologically produced from simple sugars without the production of carbon dioxide:

$$C_6H_{12}O_6 \rightarrow 2\ CH_3CH_2OH + 2\ CO_2$$

(Biological production of ethanol)

$$C_6H_{12}O_6 \rightarrow 3\ CH_3COOH$$

(Biological production of acetic acid)

Because of this, on a mass basis, the yields will be higher than in ethanol fermentation. If then, the undigested residue (mostly lignin) is used to produce hydrogen by gasification, it is ensured that more energy from the biomass will end up as liquid fuels rather than excess heat/electricity .

$$3\ CH_3COOH + 6\ H_2 \rightarrow 3\ CH_3CH_2OH + 3\ H_2O$$

(Hydrogenation of acetic acid)

$$C_6H_{12}O_6 \text{ (from cellulose)} + 6\ H_2 \text{ (from lignin)} \rightarrow 3\ CH_3CH_2OH + 3\ H_2O$$

(Overall reaction)

A more comprehensive description of the economics of each of the fuels is given on the pages alcohol fuel and ethanol fuel, more information about the economics of various systems can be found on the central page biofuel.

Stage of Development

The system has been in development since 1991, moving from the laboratory scale (10 g/day) to the pilot scale (200 lb/day) in 2001. A small demonstration-scale plant (5 ton/day) is under construction as is expected to be operational mid 2008 and a 100 ton/day demonstration plant is expected in 2009.

MICROBIAL BIODEGRADATION

Interest in the microbial biodegradation of pollutants has intensified in recent years as mankind strives to find sustainable ways to cleanup contaminated environments. These bioremediation and biotransformation methods endeavour to harness the astonishing, naturally occurring, ability of microbial xenobiotic metabolism to degrade, transform or accumulate a huge range of compounds including hydrocarbons (e.g. oil), polychlorinated biphenyls (PCBs), polyaromatic hydrocarbons (PAHs), pharmaceutical substances, radionuclides and metals. Major methodological breakthroughs in recent years have enabled detailed genomic, metagenomic, proteomic, bioinformatic and other high-throughput analyses of environmentally relevant microorganisms providing unprecedented insights into key biodegradative pathways and the ability of organisms to adapt to changing environmental conditions.

The elimination of a wide range of pollutants and wastes from the environment is an absolute requirement to promote a sustainable development of our society with low environmental impact. Biological processes play a major role in the removal of contaminants and they take advantage of the astonishing catabolic versatility of microorganisms to degrade/convert such compounds. New methodological breakthroughs in sequencing, genomics, proteomics, bioinformatics and imaging are producing vast amounts of information. In the field of Environmental Microbiology, genome-based global studies open a new era providing unprecedented *in silico* views of metabolic and regulatory networks, as well as clues to the evolution of degradation pathways and to the molecular adaptation strategies to changing environmental conditions. Functional genomic and metagenomic approaches are increasing our understanding of the relative importance of different pathways and regulatory networks to carbon flux in particular environments and for particular compounds and they will certainly accelerate the development of bioremediation technologies and biotransformation processes.

Aerobic Biodegradation of Pollutants

The burgeoning amount of bacterial genomic data provides unparalleled opportunities for understanding the genetic and molecular bases of the degradation of organic pollutants. Aromatic compounds are among the most recalcitrant of these pollutants and lessons can be learned from the recent genomic studies of *Burkholderia xenovorans* LB400 and *Rhodococcus* sp. strain RHA1, two of the largest bacterial genomes completely sequenced to date.

These studies have helped expand our understanding of bacterial catabolism, non-catabolic physiological adaptation to organic compounds, and the evolution of large bacterial genomes.

First, the metabolic pathways from phylogenetically diverse isolates are very similar with respect to overall organization.

Thus, as originally noted in pseudomonads, a large number of "peripheral aromatic" pathways funnel a range of natural and xenobiotic compounds into a restricted number of "central aromatic" pathways.

Nevertheless, these pathways are genetically organized in genus-specific fashions, as exemplified by the b-ketoadipate and Paa pathways. Comparative genomic studies further reveal that some pathways are more widespread than initially thought.

Thus, the Box and Paa pathways illustrate the prevalence of non-oxygenolytic ring-cleavage strategies in aerobic aromatic degradation processes. Functional genomic studies have been useful in establishing that even organisms harboring high numbers of homologous enzymes seem to contain few examples of true redundancy.

For example, the multiplicity of ring-cleaving dioxygenases in certain rhodococcal isolates may be attributed to the cryptic aromatic catabolism of different terpenoids and steroids. Finally, analyses have indicated that recent genetic flux appears to have played a more significant role in the evolution of some large genomes, such as LB400's, than others. However, the emerging trend is that the large gene repertoires of potent pollutant degraders such as LB400 and RHA1 have evolved principally through more ancient processes. That this is true in such phylogenetically diverse species is remarkable and further suggests the ancient origin of this catabolic capacity.

Anaerobic Biodegradation of Pollutants

Anaerobic microbial mineralization of recalcitrant organic pollutants is of great environmental significance and involves intriguing novel biochemical reactions. In particular, hydrocarbons and halogenated compounds have long been doubted to be degradable in the absence of oxygen, but the isolation of hitherto unknown anaerobic hydrocarbon-degrading and reductively dehalogenating bacteria during the last decades provided ultimate proof for these processes in nature.

Many novel biochemical reactions were discovered enabling the respective metabolic pathways, but progress in the molecular understanding of these bacteria was rather slow, since genetic systems are not readily applicable for most of them. However, with the increasing application of genomics in the field of environmental microbiology, a new and promising perspective is now at hand to obtain molecular insights into these new metabolic properties.

Several complete genome sequences were determined during the last few years from bacteria capable of anaerobic organic pollutant degradation. The ~4.7 Mb genome of the facultative denitrifying *Aromatoleum aromaticum* strain EbN1 was the first to be determined for an anaerobic hydrocarbon degrader (using toluene or ethylbenzene as substrates). The genome sequence revealed about two dozen gene clusters (including several paralogs) coding for a complex catabolic network for anaerobic and aerobic degradation of aromatic compounds.

The genome sequence forms the basis for current detailed studies on regulation of pathways and enzyme structures. Further genomes of anaerobic hydrocarbon degrading bacteria were recently completed for the iron-reducing species *Geobacter metallireducens* (accession nr. NC_007517) and the perchlorate-reducing *Dechloromonas aromatica* (accession nr. NC_007298), but these are not yet evaluated in formal publications. Complete genomes were also determined for bacteria capable of anaerobic degradation of halogenated hydrocarbons by halorespiration: the ~1.4 Mb genomes of *Dehalococcoides ethenogenes* strain 195 and *Dehalococcoides* sp. strain CBDB1 and the ~5.7 Mb genome of *Desulfitobacterium hafniense* strain Y51. Characteristic for all these bacteria is the presence of multiple paralogous genes for reductive dehalogenases, implicating a wider dehalogenating spectrum of the organisms than previously known. Moreover, genome sequences provided unprecedented insights into the evolution of reductive dehalogenation and differing strategies for niche adaptation.

Bioavailability, Chemotaxis, and Transport of Pollutants

Bioavailability, or the amount of a substance that is physiochemically accessible to microorganisms is a key factor in the efficient biodegradation of pollutants. Chemotaxis, or the directed movement of motile organisms towards or away from chemicals in the environment is an important physiological response that may contribute to effective catabolism of molecules in the environment. In addition, mechanisms for the intracellular accumulation of aromatic molecules via various transport mechanisms are also important.

Oil Biodegradation

Petroleum oil contains aromatic compounds that are toxic for most life forms. Episodic and chronic pollution of the environment by oil causes major ecological perturbations. Marine environments are especially vulnerable since oil spills of coastal regions and the open sea are poorly containable and mitigation is difficult. In addition to pollution through human activities, millions of tons of petroleum enter the marine environment every year from natural seepages. Despite its toxicity, a considerable fraction of petroleum oil entering marine systems is eliminated by the hydrocarbon-degrading activities of microbial communities, in particular by a remarkable recently discovered group of specialists, the so-called hydrocarbonoclastic bacteria (HCB) . *Alcanivorax borkumensis* was the first HCB to have its genome sequenced.

Analysis of Waste Biotreatment

Sustainable development requires the promotion of environmental management and a constant search for new technologies to treat vast quantities of wastes generated by increasing anthropogenic activities. Biotreatment, the processing of wastes using living organisms, is an environmentally friendly, relatively simple and cost-effective alternative to physico-chemical clean-up options. Confined environments, such as bioreactors], have been engineered to overcome the physical, chemical and biological limiting factors of biotreatment processes in highly controlled systems. The great versatility in the design of confined environments allows the treatment of a wide range of wastes under optimized conditions. To perform a correct assessment, it is necessary to consider various microorganisms having a variety of genomes and expressed transcripts and proteins. A great number of analyses are often required. Using traditional genomic techniques, such assessments are limited and time-consuming. However, several high-throughput techniques originally developed for medical studies can be applied to assess biotreatment in confined environments.

Metabolic Engineering and Biocatalytic Applications

The study of the fate of persistent organic chemicals in the environment has revealed a large reservoir of enzymatic reactions with a large potential in preparative organic synthesis, which has already been exploited for a number of oxygenases on pilot and even on industrial scale. Novel catalysts can be obtained from metagenomic libraries and DNA sequence based approaches.

Our increasing capabilities in adapting the catalysts to specific reactions and process requirements by rational and random mutagenesis broadens the scope for application in the fine chemical industry, but also in the field of biodegradation. In many cases, these catalysts need to be exploited in whole cell bioconversions or in fermentations, calling for system-wide approaches to understanding strain physiology and metabolism and rational approaches to the engineering of whole cells as they are increasingly put forward in the area of systems biotechnology and synthetic biology.

BIOTRANSFORMATION

Biotransformation is the chemical modification (or modifications) made by an organism on a chemical compound. If this modification ends in mineral compounds like CO_2, NH_3^+ or H_2O, the biotransformation is called mineralisation.

Drug Metabolism

The metabolism of a drug or toxin in a body is an example of a biotransformation. Typically the body deals with a foreign compound by making it more soluble, to increase the rate of its excretion through the urine. There are a number of different process that can occur,The pathways of drug metabolism can divided in to:

- phase I
- phase II

Phase I reaction

- It includes oxidative, reductive and hydrolytic reactions.
- In case of these type of reactions, the polar group is either introduced or unmasked,so the drug molecule becomes solublised and excreted.

Phase II reaction

- These reactions involve covalent attachment of small polar endogenous molecule like glucuronic acid, sulphate, glycine etc. to form highly water soluble substances.
- Thus, they are known as conjugation reaction
- The formed products have more molecular size. So, termed as synthetic reaction.

Microbial Biotransformation

Biotransformation of various pollutants is a sustainable way to clean up contaminated environments. These bioremediation and biotransformation methods harness the naturally occurring, microbial catabolic diversity to degrade, transform or accumulate a huge range of compounds including hydrocarbons (e.g. oil), polychlorinated biphenyls (PCBs), polyaromatic hydrocarbons (PAHs), pharmaceutical substances, radionuclides and metals.

Major methodological breakthroughs in recent years have enabled detailed genomic, metagenomic, proteomic, bioinformatic and other high-throughput analyses of environmentally relevant microorganisms providing unprecedented insights into biotransformation and biodegradative pathways and the ability of organisms to adapt to changing environmental conditions.

Biological processes play a major role in the removal of contaminants and pollutants from the environment. Some microorganisms possess an astonishing catabolic versatility to degrade or transform such compounds.

New methodological breakthroughs in sequencing, genomics, proteomics, bioinformatics and imaging are producing vast amounts of information.

In the field of Environmental Microbiology, genome-based global studies open a new era providing unprecedented *in silico* views of metabolic and regulatory networks, as well as clues to the evolution of biochemical pathways relevant to biotransformation and to the molecular adaptation strategies to changing environmental conditions.

Functional genomic and metagenomic approaches are increasing our understanding of the relative importance of different pathways and regulatory networks to carbon flux in particular environments and for particular compounds and they are accelerating the development of bioremediation technologies and biotransformation processes.

Also there is other approach of biotransformation called enzymatic biotransformation.

Oil Biodegradation

Petroleum oil is toxic for most life forms and episodic and chronic pollution of the environment by oil causes major ecological perturbations.

Marine environments are especially vulnerable since oil spills of coastal regions and the open sea are poorly containable and mitigation is difficult. In addition to pollution through human activities, millions of tons of petroleum enter the marine environment every year from natural seepages.

Despite its toxicity, a considerable fraction of petroleum oil entering marine systems is eliminated by the hydrocarbon-degrading activities of microbial communities, in particular by a remarkable recently discovered group of specialists, the so-called hydrocarbonoclastic bacteria (HCB).

Alcanivorax borkumensis, a paradigm of HCB and probably the most important global oil degrader, was the first to be subjected to a functional genomic analysis. This analysis has yielded important new insights into its capacity for:

(i) n-alkane degradation including metabolism, biosurfactant production and biofilm formation,

(ii) scavenging of nutrients and cofactors in the oligotrophic marine environment, as well as

(iii) coping with various habitat-specific stresses.

The understanding thereby gained constitutes a significant advance in efforts towards the design of new knowledge-based strategies for the mitigation of ecological damage caused by oil pollution of marine habitats.

HCB also have potential biotechnological applications in the areas of bioplastics and biocatalysis.

Metabolic Engineering and Biocatalytic Applications

The study of the fate of persistent organic chemicals in the environment has revealed a large reservoir of enzymatic reactions with a large potential in preparative organic synthesis, which has already been exploited for a number of oxygenases on pilot and even on industrial scale. Novel catalysts can be obtained from metagenomic libraries and DNA sequence based approaches. Our increasing capabilities in adapting the catalysts to specific reactions and process requirements by rational and random mutagenesis broadens the scope for application in the fine chemical industry, but also in the field of biodegradation. In many cases, these catalysts need to be exploited in whole cell bioconversions or in fermentations, calling for system-wide approaches to understanding strain physiology and metabolism and rational approaches to the engineering of whole cells as they are increasingly put forward in the area of systems biotechnology and synthetic biology.[I]

XENOBIOTIC METABOLISM

Xenobiotic metabolism is the set of metabolic pathways that modify the chemical structure of xenobiotics, which are compounds foreign to an organism's normal biochemistry, such as drugs and poisons. These pathways are a form of biotransformation present in all major groups of organisms, and are considered to be of ancient origin. These reactions often act to detoxify poisonous compounds; however, in some cases, the intermediates in xenobiotic metabolism can themselves be the cause of toxic effects.

Xenobiotic metabolism is divided into three phases. In phase I, enzymes such as cytochrome P450 oxidases introduce reactive or polar groups into xenobiotics. These modified compounds are then conjugated to polar compounds in phase II reactions. These reactions are catalysed by transferase enzymes such as glutathione S-transferases. Finally, in phase III, the conjugated xenobiotics may be further processed, before being recognised by efflux transporters and pumped out of cells.

The reactions in these pathways are of particular interest in medicine as part of drug metabolism and as a factor contributing to multidrug resistance in infectious diseases and cancer chemotherapy. The actions of some drugs as substrates or inhibitors of enzymes involved in xenobiotic metabolism are a common reason for hazardous drug interactions. These pathways are also important in environmental science, with the xenobiotic metabolism of microorganisms determining whether a pollutant will be broken down during bioremediation, or persist in the environment.

Permeability Barriers and Detoxification

That the exact compounds an organism is exposed to will be largely unpredictable, and may differ widely over time, is a major characteristic of xenobiotic toxic stress. The major challenge faced by xenobiotic detoxification systems is that they must be able to remove the almost-limitless number of xenobiotic compounds from the complex mixture of chemicals involved in normal metabolism. The solution that has evolved to address this problem is an elegant combination of physical barriers and low-specificity enzymatic systems.

All organisms use cell membranes as hydrophobic permeability barriers to control access to their internal environment. Polar compounds cannot diffuse across these cell membranes, and the uptake of useful molecules is mediated through transport proteins that specifically select substrates from the extracellular mixture. This selective uptake means that most hydrophilic molecules cannot enter cells, since they are not recognised by any specific transporters. In contrast, the diffusion of hydrophobic compounds across these barriers cannot be controlled, and organisms, therefore, cannot exclude lipid-soluble xenobiotics using membrane barriers.

However, the existence of a permeability barrier means that organisms were able to evolve detoxification systems that exploit the hydrophobicity common to membrane-permeable xenobiotics. These systems therefore solve the specificity problem by possessing such broad substrate specificities that they metabolise almost any non-polar compound. Useful metabolites are excluded since they are polar, and in general contain one or more charged groups.

Phases of detoxification

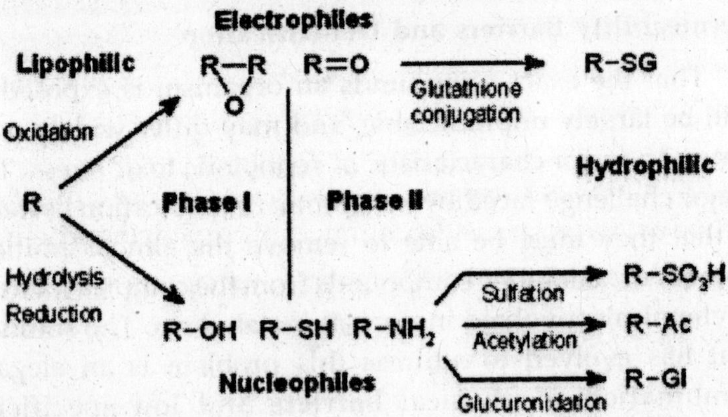

Phases I and II of the metabolism of a lipophilic xenobiotic.

The detoxification of the reactive by-products of normal metabolism cannot be achieved by the systems outlined above, because these species are derived from normal cellular constituents and usually share their polar characteristics.

However, since these compounds are few in number, specific enzymes can recognize and remove them. Examples of these specific detoxification systems are the glyoxalase system, which removes the reactive aldehyde methylglyoxal, and the various antioxidant systems that eliminate reactive oxygen species.

The metabolism of xenobiotics is often divided into three phases: modification, conjugation, and excretion. These reactions act in concert to detoxify xenobiotics and remove them from cells.

Phase I - modification

In phase I, a variety of enzymes acts to introduce reactive and polar groups into their substrates. One of the most common modifications is hydroxylation catalysed by the cytochrome P-450-dependent mixed-function oxidase system. These enzyme complexes act to incorporate an atom of oxygen into nonactivated hydrocarbons, which can result in either the introduction of hydroxyl groups or N-, O- and S-dealkylation of substrates. The reaction mechanism of the P-450 oxidases proceeds through the reduction of cytochrome-bound oxygen and the generation of a highly-reactive oxyferryl species, according to the following scheme:

$$NADPH + H^+ + RH \rightarrow NADP^+ + H_2O + ROH$$

Phase II - conjugation

In subsequent phase II reactions, these activated xenobiotic metabolites are conjugated with charged species such as glutathione (GSH), sulfate, glycine, or glucuronic acid. These reactions are catalysed by a large group of broad-specificity transferases, which in combination can metabolise almost any hydrophobic compound that contains nucleophilic or electrophilic groups. One of the most important of these groups are the glutathione S-transferases (GSTs). The addition of large anionic groups (such as GSH) detoxifies reactive electrophiles and produces more polar metabolites that cannot diffuse across membranes, and may, therefore, be actively transported.

Phase III - further modification and excretion

After phase II reactions, the xenobiotic conjugates may be further metabolised. A common example is the processing of glutathione conjugates to acetylcysteine (mercapturic acid) conjugates. Here, the ã-glutamate and glycine residues in the glutathione molecule are removed by Gamma-glutamyl transpeptidase and dipeptidases. In the final step, the cystine residue in the conjugate is acetylated.

Conjugates and their metabolites can be excreted from cells in phase III of their metabolism, with the anionic groups acting as affinity tags for a variety of membrane transporters of the multidrug resistance protein (MRP) family. These proteins are members of the family of ATP-binding cassette transporters and can catalyse the ATP-dependent transport of a huge variety of hydrophobic anions, and thus act to remove phase II products to the extracellular medium, where they may be further metabolised or excreted.

Endogenous Toxins

The detoxification of endogenous reactive metabolites such as peroxides and reactive aldehydes often cannot be achieved by the system described above. This is the result of these species' being derived from normal cellular constituents and usually sharing their polar characteristics. However, since these compounds are few in number, it is possible for enzymatic systems to utilize specific molecular recognition to recognize and remove them. The similarity of these molecules to useful metabolites therefore means that different detoxification enzymes are usually required for the metabolism of each group of endogenous toxins. Examples of these specific detoxification systems are the glyoxalase system, which acts to dispose of the reactive aldehyde methylglyoxal, and the various antioxidant systems that remove reactive oxygen species.

History

Studies on how people transform the substances that they ingest began in the mid-nineteenth century, with chemists discovering that organic chemicals such as benzaldehyde could be oxidized and conjugated to amino acids in the human body. During the remainder of the nineteenth century, several other basic detoxification reactions were discovered, such as methylation, acetylation, and sulfonation.

In the early twentieth century, work moved on to the investigation of the enzymes and pathways that were responsible for the production of these metabolites. This field became defined as a separate area of study with the publication by Richard Williams of the book *Detoxication mechanisms* in 1947. This modern biochemical research resulted in the identification of glutathione *S*-transferases in 1961, followed by the discovery of cytochrome P450s in 1962, and the realization of their central role in xenobiotic metabolism in 1963.

BIODEGRADABILITY PREDICTION

Biodegradability prediction is biologically-inspired computing and attempts to predict biodegradability of anthropogenic materials in the environment. Demand for biodegradability prediction is expected to increase with governments stepping up environmental regulations (see, for instance, testing for bioaccumulation in the REACH proposal).

Example:

- Development of quantitative structure-activity relationship (QSARs) for biodegradation, for instance, biochemical oxygen demand for chemicals released into the environment with the aid of machine learning and other artificial intelligence methods.
- The University of Minnesota Biocatalysis and Biodegradation Database (UM-BBD), which contains information on microbial biocatalytic reactions and biodegradation pathways for primarily xenobiotic, chemical compounds. One of its many features allows the prediction of microbial catabolic reactions using substructure searching, a rule-base, and atom-to-atom mapping.

Composting

Composting is the aerobic decomposition of biodegradable organic matter, producing compost. (Or in a simpler form: Composting is the decaying of food, mostly vegetables or manure.)

The decomposition is performed primarily by facultative and obligate aerobic bacteria, yeasts and fungi, helped in the cooler initial and ending phases by a number of larger organisms, such as ils, and other families representing ants, nematodes and oligochaete worms.

Composting can be divided into home composting and industrial composting. Essentially the same biological processes are involved in both scales of composting, however techniques and different factors must be taken into account.

Importance

Composting recycles or "downcycles" organic household and yard waste and manures into an extremely useful humus-like, soil end-product called compost. Examples are fruits, vegetables and yard clippings. Ultimately this permits the return of needed organic matter and nutrients into the foodchain and reduces the amount of "green" waste going into landfills. Composting is widely believed to speed up the natural process of decomposition appreciably as a result of the raised temperatures that often accompany it.

The elevated heat results from exothermic processes, and the heat in turn reduces the generational time of microorganisms and thereby speeds the energy and nutrient exchanges taking place. It is a popular misconception that composting is a "controlled" process; if the right environmental circumstances are present the process virtually runs itself. Hence a popular expression, "compost happens". It is nonetheless very necessary to provide as optimal circumstances as possible for large amounts of organic waste to break down properly. This is especially so when it is accompanied by heating, since at elevated temperatures oxygen within the piles is consumed more rapidly, and if not controlled, will lead to malodor.

Decomposition similar to composting occurs throughout nature as garbage dissolves in the absence of all the conditions that modern composters talk about; however, the process can be slow. For example, in the forest bark, wood and leaves break down into humus over 3-7 years. In restricted environments, for example, vegetables in a plastic trash container, decomposition with a lack of air encourages growth of anaerobic microbes, which produce disagreeable odors. Another form of degradation practiced deliberately in absence of oxygen is called anaerobic digestion- an increasingly popular companion to composting as it enables capture of residual energy in the form of biogas, whereas composting releases the majority of bound carbon-energy as excess heat (which helps sanitize the material) as well as copious amounts of biogenic CO_2 to the atmosphere.

It is important to distinguish between terms such as "biodegradable", "compostable", and "compost-compatible".

- A biodegradable material is capable of being broken down completely under the action of microorganisms into carbon dioxide, water and biomass. It may take a very long time for a material to biodegrade depending on its environment (e.g. hardwood in an arid area), but it ultimately breaks down completely.
- A compostable material biodegrades substantially under composting conditions, into carbon dioxide, methane, water and compost biomass. Compost biomass refers to the portion of the material that is metabolized by the microorganisms and which is incorporated into the cellular structure of the organisms or converted into humic acids etc. Compost biomass residues from a compostable material are fully biodegradable. "Compostable" is thus a subset of "biodegradable". The size of the material is a factor in determining compostability because it affects the rate of degradation. Large pieces of hardwood may not be compostable under a specific set of composting conditions, whereas sawdust of the same type of wood may be.
- A compost-compatible material does not have to be compostable or even biodegradable. It may biodegrade too slowly to be compostable itself, or it may not biodegrade at all. However, it is not readily

distinguishable from the compost on a macroscopic scale and does not have a deleterious effect on the compost (e.g. it is not a biocide). Compost-compatible materials are generally inert and are present in compost at relatively low levels. Examples of compost-compatible materials include sand particles and inert particles of plastic.

Reducing Waste

Although composting has historically focussed on creating garden-ready soil, it is becoming more important as a tool is reducing solid waste. More than 60 percent of household waste in the United States is recyclable or compostable. But Americans only compost 8 percent of their waste.

Surveys have shown that the #1 reason Americans don't compost their waste is because they feel the process is complicated, time-consuming or requires special equipment. However, especially in rural areas, much of the solid waste could be removed from the waste stream by promoting "extremely passive composting" where consumers simply discard their yard waste and kitchen scraps on their own land, regardless of whether the material is ever re-used as "compost".

Materials

Many different materials are suitable for composting organisms. Composters often refer to "C:N" requirements; some materials contain high amounts of carbon in the form of cellulose which the bacteria need for their energy. Other materials contain nitrogen in the form of protein, which provide nutrients for the energy exchanges. It would however be an over-simplification to describe composting as about carbon and nitrogen, as is often portrayed in popular literature. Elemental carbon - such as charcoal - is not compostable nor is a pure form of nitrogen, even in combination with carbon. Not only this, but a great variety of man-made, carbon-containing products, including many textiles and polyethylene, are not compostable - hence the push for biodegradable plastics.

Suitable ingredients with relatively high carbon content include:

- Dry, straw-type material, such as cereal straws
- Autumn leaves
- Sawdust and wood chips
- paper and cardboard (such as corrugated cardboard or newsprint with soy-based inks)

Ingredients with relatively high nitrogen content include:

- Green plant material (fresh or wilted) such as crop residues, hay, grass clippings, weeds
- Manure of poultry and herbivorous animals such as horses, cows and llamas
- Fruit and vegetable trimmings

The most efficient composting occurs by seeking to obtain an initial C:N mix of 25/30 by dry chemical weight. Grass clippings have an average ratio of 10-19 to 1 and dry autumn leaves from 55-100 to 1. Mixing equal parts by volume approximates the ideal range.

Poultry manure provides much nitrogen but with a ratio to carbon that is imbalanced. If composted alone, this results in excessive N-loss in the form of ammonia - and some odor. Horse manure provides a good mix of both, although in modern stables, so much bedding may be used as to make the mix too carbonaceous. For home-scale composting, mixing the materials as they are added increases the rate of decomposition, but it can be easier to place the materials in alternating layers, approximately 15 cm (6 in) thick, to help estimate the quantities. Keeping carbon and nitrogen sources separated in the pile can slow down the process, but decomposition will still occur.

Some people put special materials and activators into their compost. A light dusting of agricultural lime (not on animal manure layers) can curb excessive acidity, especially with food waste.

Seaweed meal provides a ready source of trace elements. Finely pulverized rock (rock flour or rock dust) can also provide minerals, while clay and leached rock dust are poor in trace minerals.

Composting in the form of bioremediation can break down petroleum hydrocarbons, TNT and a variety of toxic compounds. This is the bacterial and in some cases fungal content of the compost, that possess the enzymatic properties to de-polymerize the complex man-made molecules. In other words, there is nothing about the composting process *per se* that adds or detracts from this, unless as noted above, by warming, to increase the metabolic rate of the constituent organisms.

Some materials are best left to high-rate, a thermophilic composting system, as they decompose slower, attract vermin and require higher temperatures to kill pathogens than backyard composting provides.

These materials include meat, dairy products, eggs, restaurant grease, cooking oil, manure and bedding of non-herbivores, and residuals from the treatment of wastewater and drinking water. Meat and dairy products can be recycled using bokashi, a fermentation method. Human waste can be composted by industrial, high-heat methods and also composting toilets, even though most composting toilets do not allow for the thermophilic decomposition believed to be necessary rapid kill of pathogens, such as Salmonella This is not a problem, however, since composting toilets also incorporate the essential element of time required to reduce available substrate on which pathogens can feed, while increasing the growth of competing microbes.

If these high temperatures are reached, the resulting compost can be safely used as a fertilizer for food crops and even directly edible crops provided it is not illegal in the regions where the sludge is applied. Careful filtration of the compost also prevents contamination.

Approaches

There are two major approaches to composting: active and passive. These terms are somewhat of a misnomer since both active and passive composts can attain high heating, which increases the rate of biochemical processes. But the terms active and passive are appropriate descriptions for the nature of human intervention used.

Active

Active (hot) composting is composting at close to ideal conditions, allowing aerobic bacteria to thrive. Aerobic bacteria break down material faster and produce less odor and fewer pathogens and destructive greenhouse gases than anaerobic bacteria. Commercial-grade composting operations actively control the composting conditions such as the carbon-to-nitrogen ratio. For backyard composters, the charts of carbon and nitrogen ratios in various ingredients and the calculations required to get the ideal mixture can be intimidating, so many rules of thumb exist for approximating it.

Pasteurisation is a misnomer in composting, as no compost will become truly sterilized by high temperatures alone. Rather, in a very hot compost where the temperature exceeds 55 °C (130 °F) for several days, the ability of organisms to survive is greatly compromised. Nevertheless, there are many organisms in nature that can survive extreme temperatures, including the group of pathogenic Clostridium, and so no compost is completely safe. To achieve the elevated temperatures, the compost bin must be kept warm, insulated and damp.

Aerated Composting is an efficient form of composting from the chemical point of view as it produces ultimately only energy in the form of waste heat and CO_2 and H_2O. With aerated composting, fresh air (i.e. oxygen) is introduced throughout the mix of materials using any appropriate mechanism.

The air stimulates the microorganisms that are already in the mix, and their by-product is heat. In a properly operated compost system, pile temperatures are sufficient to stabilize the raw material, and the oxygen-rich conditions within the core of the pile eliminate offensive odors. High temperatures also destroy fly larvae and weed seeds, yielding a safe, high-quality finished product.

Finally, aeration expedites the composting process through the mechanism of heating insofar as the elevated heat will drive biochemical processes faster, so that a finished product can be rendered in 60 to 120 days. Aerated compost is an excellent source of macro- and micro-nutrients as well as stable organic matter, all of which support healthy plant growth. In addition, the micro-organisms in compost aid in the suppression of plant pathogens. Finally, compost retains water extremely well resulting in improved drought resistance, a longer growing season, and reduced soil erosion.

In Thailand, May 2008, this system has been used by farmer groups for more than 445 sites. The process needs only 30 days to be finished and 10 metric tons of compost is obtained each time. A blower (15 inch squirrel-caged blower with 3 hp motor) is needed to force the air through 10 piles of compost twice a day and 15 minutes each. The raw materials consist of agricultural wastes and animal manure in the ratio of 3 : 1 by volume. For more information please visit www.compost.mju.ac.th/eng

Passive

Passive composting is composting in which the level of physical intervention is kept to a minimum, and often as a result the temperatures never reach much above 30°C (86 °F). It is slower but is the more common type of composting in most domestic garden compost bins. Such composting systems may be either enclosed (home container composting, industrial in-vessel composting) or in exposed piles (industrial windrow composting). Kitchen scraps are put in the garden compost bin and left untended. This scrap bin can have a very high water content which reduces aeration, and so becomes odorous. To improve drainage and airflow, a gardener can mix in wood chips, small pieces of bark, leaves or twigs, or make physical holes through the pile.

Natural

An unusual form of natural composting in nature is seen in the case of the mound-builders (megapodes) of eastern Indonesia, New Guinea, and Australia as well as in the case of bowerbirds of New Guinea and Australia. These

Megapodes are fowl-sized birds famous for building nests in the form of huge compost heaps containing leaf litter, in which they incubate their eggs. The birds work constantly to maintain the correct, almost exact, incubation temperatures, by adding and removing leaves from the compost pile. In effect, this teaches us that thermophilic high-temperature composting is not man-made.

Home Composting

Home composters use a range of techniques, varying from extremely passive (throw everything in a pile and leave it for a year or two) to extremely active (monitor the temperature, turn the pile regularly, and adjust the ingredients over time). Some composters use mineral powders to absorb smells, although a well-maintained pile seldom has bad odors. It is usually located in the back garden.

Moisture and Heat

An effective compost pile is about as damp as a well wrung-out sponge. This provides the moisture that all life requires. Microorganisms vary by their ideal temperature and the heat they generate as they digest. Mesophilic bacteria survive best at temperatures of 20 to 44 °C (70 to 120 °F). thermophilic (heat-surviving) bacteria grow optimally at around 55°C (130 °F), and can attain the fastest decomposition, since metabolic processes proceed more rapidly under higher temperatures. Elevated temperature is also preferred since it causes the most rapid pathogen reduction, and is more destructive of weed seeds. To minimally achieve it, the heap should be about 1 m (3 ft) wide, 1 m (3 ft) tall, and as long as is practicable. This provides enough insulating mass to build up heat but also allows aeration. The center of the pile heats up the most.

If the pile does not heat up, common reasons include that:

- The heap is too wet, limiting the oxygen which bacteria require
- The heap is too dry for the bacteria to survive and reproduce
- There is insufficient protein (nitrogen-rich material)

The necessary material should be added, or the pile should be turned to aerate it and bring the outer layers inside and vice versa. You should add water at this time to help keep the pile damp. One guideline is to turn the pile when the high temperature has begun to drop, indicating that the food source for the fastest-acting bacteria (in the center of the pile) has been largely consumed. When turning the pile does not cause a temperature rise, it brings no further advantage. When all the material has turned into dark brown crumbly matter, it is ready to use.

Worm Composting

Worm composting or vermicomposting is a method of composting using Red Wiggler worms in a container. Food waste and moistened bedding are added, and the worms and micro-organisms eventually convert them to rich compost. The worms excrete a soil-nutrient material called worm castings. Worm composting can be done indoors, allowing year-round composting, and providing apartment dwellers with a means of composting.

Worms are low in the food chain, and so are critical to healthy soil. This is why farmers have historically wanted healthy worm populations to live in their fields.

The nutrients and micro-organisms can be concentrated in liquid form called worm tea, made by running distilled water through worm castings. When it is poured into the soil, the microorganisms multiply, creating a healthy growing environment for plants.

Industrial Composting

Industrial composting systems are increasingly being installed as a waste management alternative to landfills, along with other advanced waste processing systems. Industrial composting or anaerobic digestion combined with mechanical sorting of mixed waste streams is called mechanical biological treatment increasingly used in Europe due to stringent new regulations controlling the amount of organic matter allowed in landfills. Treating biodegradable waste before it enters a landfill reduces global warming from fugitive methane; untreated waste breaks down anaerobically in a landfill, producing landfill gas that contains methane, a greenhouse gas even more potent than carbon dioxide.

Most commercial and industrial composting operations use active composting techniques. These ensure that the process does not get out of control especially with the high through-put demand imposed by contracted, incoming waste. This means that as short as possible a processing time must be maintained to keep the facility properly functioning (see compost windrow turner). Partly for this reason composters have declined to support compost maturity standards if it would increase the required holding time. The greatest amount of technological control of composting is seen in systems using an enclosed vessel and controlling its temperature, air flow, moisture and other parameters. See In-vessel composting (indoor composting).

Large-scale composting systems are used by many urban centers around the world. Co-composting is a technique which combines solid waste with de-watered biosolids, which originated in the 1960s and has fallen somewhat out of favor due to difficulties controlling inert and plastic contamination from MSW. In Europe, mixed waste composting is virtually illegal. The world's largest MSW co-composter is the Edmonton Composting Facility in Edmonton in Alberta, Canada, which turns 220,000 tonnes of residential solid waste and 22,500 dry tonnes of biosolids per year into 80,000 tonnes of compost. The facility is 38,690 square metres (416,500 ft^2) large (equivalent to 4½ Canadian football fields), and the aeration building alone is the largest stainless steel building in North America, the size of 14 NHL rinks.

In broadest sense, industrial microbiology, is concerned with all aspects of business that relate to microbiology. In a more restricted sense, industrial microbiology is concerned with

(i) employing microorganisms to produce a desired product, and with

(ii) preventing microbes from diminishing the economic value of various products. This dual purpose is clearly seen in the food industry, a major area of industrial microbiology discussed in separate chapter on "Microbiology of foods, Milk and Dairy Products".

Various commercial products of economic value made by microbes are

(i) medicines i.e. pharmaceuticals, including antibiotics, steroids, human protein, vaccines, and vitamins,

(ii) organic acids,

(iii) amino acids,

(vi) enzymes,

(v) alcohols,

(vi) organic solvents and

(vii) synthetic fuels. In addition to these, quite recently potential of microbes could also be realised in

(viii) recovery of metals from ores through bioleaching,

(ix) recovery of petrol, and

(x) single all protein production.

The term fermentation in industrial microbiology is used in a wider sense to include any chemical transformation of organic compounds carried out by using microbes and their enzymes.

Production methods in industrial microbiology bring together the raw materials (substrates), microorganisms (specific strains or microbial enzymes) and a controlled favourable environment (created in a ferment or) to produce the desired substance.

The essence of an industrial process is to combine the right organism, an in expensive substrate, and the proper environment to produce high yields of a desired product

Microbes possess a wealth of metabolic equipment that brings about diverse chemical transformations. This characteristic of microbes could be exploited in obtaining some valuable products of daily use. The cheap raw materials available in nature as a waste, may be converted into useful commercial products by the activity of microbes. Microbes thus serve a dual purpose. Firstly, they are good agents of disposal of these wastes, and secondly the resultant end products of their breakdown are seful commercial products.

Industrial microbiology is an important area of applied microbiology. It refers to the use of microorganisms in commercial enterprise and . Cheap raw materials are converted to valuable products through the metabolism of microbes. Microbes for this purpose could be exploited in different ways. For instance this includes

(i) synthesis of fermentation products as acids, alcohols or other organic compounds

(ii) transformation of one compound into another desired type,

(iii) the production of enzymes, antibiotics, or insecticides, or

(iv) the use of microbes themselves as food.

Ethyl Alcohol Production by Microbiological Processes - Production of ethyl alcohol from sugary materials is one of the oldest known microbiological processes. Alcohol is an important solvent and raw material used in a variety of chemical industries. Although to day industrial alcohol is also produced synthetically from ethylene, production of alcohol by fermentation of cheap sugary materials such as molasses by yeast, is still an important industry.

For ethyl alcohol production, selected strains of Saccharomyces cerevisiae are employed since all the strains are not equally efficient. The alcohol tolerance and sugar tolerance are important criteria used in the selection of yeast strains. Strains tolerant to high sugar and alcohol concentration are desired. The raw material generally used is either crude cane molasses or best molasses which contain about 50 per cent fermentable sugars

The production process involves the dilution of molasses to a suitable sugar concentration (15-16 per cent sugars), addition of a small quantity of nitrogen source (urea, ammonium sulfate or ammonium phosphate), adjustment of pH to about 5.0, and the, addition of an actively growing yeast culture.

The fermentation is .carried out in big deep tanks of steel or stainless steel. The fermentation is allowed to continue for about 24-36 hours at 25°C-30°C after which the cells are allowed to settle. The fermented mash is then distilled and passed through rectifying columns to recover ethyl alcohol.

A large amount of carbon dioxide is also produced during the fermentation which is purified and compressed. The yield of ethyl alcohol is about 50 per cent of the fermentable sugar concentration. Further purification of the ethyl alcohol is done by fractional distillation. In Some distilleries, the yeast is recovered and used as animal feed while in most, it is discarded into the effluents, a procedure that is very undesirable

In recent years because of the possibility of using ethyl alcohol as a fuel supplement and a chemical feed stock, there is increased interest in increasing production but at a cheaper and economical rate. For this, a variety of improvements in the traditional batch fermentation have been described in literature. Among these, the one that has attracted attention is the cell recycle technique which does not involve much additional expenditure.

Basically, the technique involves the reuse of cell mass that is produced during the fermentation. It has been found that by doing so, about 5-10 per cent of the substrate which would have been otherwise used for cell growth is saved ill addition to a great saving in the cost of inoculum and time. By using recycling technology, fermentation time has been drastically reduced from 24-36 hours in a batch fermentation to as low as 5-6 hours.

3

The Evaluation of Mixtures of Yeast and Potato Extracts in Growth Media for Biomass Production of Lactic Cultures

The effectiveness of yeast extracts (YE) and potato extracts (PE) to promote growth of seven lactic cultures was evaluated by automated spectrophotometry (AS). Two aspects of the growth curve were analysed: (1) maximum biomass obtained (using OD_{max}) and (2) highest specific growth rate ($\grave{\imath}_{max}$).

Eleven lots from the same PE-manufacturing process were examined for lot-to-lot variability. The OD_{max} values of three of the seven strains were significantly affected by lot source, but $\grave{\imath}_{max}$ was not significantly affected.

The growth of bacteria was systematically lower in base medium containing 100% PE than in base medium containing 100% YE for both OD_{max} or $\grave{\imath}_{max}$ data, which could be related to the lower content in nitrogen-based compounds in PE.

In AS assays, highest OD values for *Lactobacillus casei* EQ28, *Lactobacillus rhamnosus* R-011, *Lactobacillus plantarum* EQ12, and *Streptococcus thermophilus* R-083 were obtained with a mixture of PE and YE.

Fermentations (2 L) were also carried out to determine the accuracy of AS to predict biomass levels obtained under fermentation trials. In these fermentations, replacement of 50% YE with PE was shown to enable good growth of *S. thermophilus*. With *L. rhamnosus* R-011, a high correlation (R^2 = 0.95) was found between OD_{max} data obtained in the AS assays and that of the 2-L bioreactor when the same growth medium was used for both series of fermentations. However, AS was not as efficient when industrial media were used for the bioreactor assays. The relationship was still good for OD_{max} between AS data and that of the bioreactor data with *L. rhamnosus* R-011 in industrial LBS medium (R^2 = 0.87), but was very poor with the *S. thermophilus* R-083 on Rosell #43 industrial medium (R^2 = 0.33). Since PE cost 40% less than YE, there are strong economic advantages in considering such a partial replacement of YE by PE.

INTRODUCTION

Lactic acid bacteria (LAB) are recognized as being fastid-ious in their nutritional requirements, and variations in nutritional requirements occur between species (Marshall and Law 1984) and even strains of the same species (Champagne et al. *1999)*. They thus constitute good tools to study the growth-promoting value of media ingredients as well as variability between production lots of a given product (Potvin et al. 1997).

Many ingredients are found in laboratory media (MRS, Elliker, or M17) or commercial starter media that are des-tined for LAB. Amino acids and vitamins are principally added through peptones, meat extracts, or yeast extracts (YE). A number of studies have shown the growth-promoting properties of YE on microorganisms (Bibal et al. 1989; Jensen and Hamer 1993; Milton et al. 1991; Ibrahim and

Bezkorovainy 1994; Olmos-Dichara et al. 1997), and they are the ingredients of choice in many media. Francou (1986) has shown that lactic yeast autolysate may be used instead of YE. Meat extracts and meat-derived peptones are not used very much on a large scale because of costs, and lately they have been a health concern. Plant extracts are generally more economical than meat extracts, and there are reports of their use with LAB (e.g., tomato juice (Difco Lab-oratories 1984), corn steep liquor (Amrane and Prigent 1993; Cox and MacBean 1977), and malt extracts (Vahvaselka and Linko 1987)).

Table 1. Source of cultures and conditions used for the preparation of the various inocula.

		Inoculum		
Strain	Source*	Medium†	Temp. (°C)	*tf* (h)‡
Lactobacillus acidophilus R-052	Rosell	MRS	37	6.5
Lactobacillus casei EQ28	Equipharm	MRS	37	6.5
Lactobacillus plantarum EQ12	Equipharm	MRS	30	8
Lactobacillus rhamnosus R-011	Rosell	MRS	37	11
Lactococcus lactis ssp. *lactis* R-058	Rosell	M17	30	16
Pediococcus pentosaceus EQ44	Equipharm	MRS	37	6.5
Streptococcus thermophilus R-083	Rosell	M17	37	5

*Subsidiaries of Lallemand Inc. (Montreal, Canada).
†MRS and M17 broth were from BDH (Toronto, Canada).
‡Fermentation time, which represents the time at which the stationary phase is reached.

There is always a need for novel, inexpensive media ingredients, and this study exam-ines the effectiveness of a potato extract (PE) novel to LAB (FNI-400; Lallemand Inc., Montr6al, Que.). PE is another product of plant origin, which has traditionally been used in growth media for yeasts (Admassu et al. 1984; Pasari et al. 1987) or molds (Chrzanowska et al. 1993; El-Banna and Leistner 1988; Murali et al. 1987). There are reports of the use of PE in media designed for bacteria (Kajimura and Kaneda 1996; Ulezlo et al. 1987), but we have found no published literature on the usefulness of PE in media de-signed for lactic cultures.

In traditional experimental plans, the nitrogen sources were studied one at a time. However, commercial media of-ten contain two nitrogen sources, and the study of multiple components is preferable to optimize growth. YE are recog-nized as being among the best nitrogen sources for LAB (Amrane and Prigent 1993; Chiarini et al. 1992; Lee et al. 1991; Aeschlimann and von Stockar 1990). Although less expensive than many peptones, YE nevertheless constitute a significant portion of media costs (Neklyudov et al. 1993; Chiarini et al. 1992), and their partial replacement with lower costing sources (some PE cost 40% less than YE) is of commercial interest. In this study, mixtures of YE and PE were added to a base medium in the aim of partially substi-tuting YE.

The optimization of growth media with multiple components is a tedious process when using fermentors or flasks; ingredients and their concentrations must be changed, which requires a great number of assays and considerable person-nel time. To answer these problems, automated spectro-photometry (AS) has been used for the determination of the growth-promoting value of YE with LAB (Potvin et al. 1997; Champagne et al. *1999)* and enables the simultaneous evaluation of up to 20 media. Results from these studies have shown a good correlation between the AS data and flask assays. However, little information has been obtained on the correlation between AS data and fermentations con-ducted under pH control, particularly when the commercial media contain turbid ingredients that cannot be studied with AS.

The aim of this study is to evaluate a novel PE for LAB biomass production, in partial replacement of YE, and to further evaluate the effectiveness of AS in selecting nitrogen sources for media formulation.

Materials and Methods

Organisms and Growth Conditions

The sources of the various lactic cultures are presented in Table 1. Stock cultures were prepared by mixing 10 mL of fresh cultures with 25 mL of 20% skim milk and 25 mL of a 20% glycerol solution (glycerol and milk were sterilized separately). The milk and glycerol cell suspensions were di-vided into 1-mL fractions, added to sterile cryovials (Nalgene, Rochester, N.Y., U.S.A.), and stored at –70°C un-til used. The inocula were prepared by adding 1 mL of thawed stock culture to the appropriate growth medium and incubating in precise conditions (Table 1). Incubations were carried out in a programmable water bath (Cole Palmer, Chi-cago, Ill., U.S.A.), and the temperatures were brought to 4°C when the desired incubation period was reached.

Yeast or Potato Extracts

Bakers' YE used in this study were obtained from BioSpringer (BioSpringer, Maisons-Alfort, France) and Lallemand Inc., while brewers' YE (X-803) were obtained by Lallemand Inc. For identification purposes, they were coded YE-BS, YE-LA, and YE-X803, respectively. The PE (FNI-400) was also supplied by Lallemand Inc. The a-amino nitrogen of PE extracts was determined by titration follow-ing reaction with formaldehyde (USP 1985). The 5% PE so-lutions were adjusted at pH 7.0 with 0.1 M NaOH or 0.1 M HCl. Total nitrogen determination was done using a FP-428 LECO apparatus (LECO Corporation, Saint-Joseph, Mich.), operated under the following conditions: sample size, 150 mg; oxidation furnace temperature, 900°C; oxidation standby temperature, 650°C; purge cycles, 3; minimum timeout, 30 s; comparator level, 1.00; loop select low range; flow constants at high; gases, oxygen 99.99% and helium 99.99%. The calibration standard was composed of 150 mg EDTA (No. 502-092) (LECO Corporation). Starch content of PE was evaluated using AOAC (1996) method 996.11.

Automated Spectrophotometry (AS) Assays

An AS instrument (Bioscreen C, Labsystems Corp., Hel-sinki, Finland) was used to assess the effect of YE and PE concentrations on strain growth, with the method described by Champagne et al. *(1999)*. The base medium (ASm) used for the growth of lactobacilli and *Pediococcus pentosaceus* strains had the following composition per litre of medium: glucose, 20 g; K_2HPO_4, 3 g; KH_2PO_4, 3 g; sodium citrate, 3 g; Tween 80, 1 g; $MgSO_4 \cdot 7H_2O$, 0.2 g; $MnSO_4 \cdot 2H_2O$, 0.05 g, and PE and (or) YE at set concentrations. Glucose and magnesium sulfate were mixed together and sterilized for 15 min at 121°C separately from the remaining ingredi-ents. For *Lactococcus lactis* and *Streptococcus thermophilus*, ASm was the same except that glucose was substituted by lactose.

In the assays aimed at evaluating mixtures of YE and PE on growth of *Lactobacillus acidophilus* R-052, 0.05% cysteine hydrochloride and 0.1% ascorbic acid were added to ASm. The ascorbic acid solution was filtered through a 0.22- m filter (Millipore, Bedford, Mass. U.S.A.) and added after sterilization of the growth medium. The data that justi-fied the addition of these compounds to the growth medium are presented in the Results and discussion section.

Sterile media containing various YE or PE were prepared in test tubes and inoculated at 1% (v/v) from a fresh culture, and 350- L samples were added aseptically to each micro-well. Microplates were incubated at 37°C for 24 h for the lactobacilli, *P. pentosaceus* EQ44 and *S. thermophilus* R-

083 and at 30°C for the strains *Lactococcus lactis* R-058 and *Lactobacillus plantarum* EQ12. The absorbency at 600 nm of each microwell was automatically recorded every 15 min. Prior to each reading, the plates were shaken at low intensity for 5 s. The various media had different initial optical den-sity (OD) values. The initial OD of the medium was sub-tracted from the OD values obtained subsequently during the fermentation. Therefore, values reported are increases in OD.

Previous studies have shown that the appropriate concen-tration of YE in the base medium, used to evaluate the bio-logical value of various YE, must be defined for each strain (Potvin et al. 1997; Champagne et al. *1999).*

Preliminary as-says were thus carried out using the method of Champagne et al. *(1999)* to determine the appropriate concentrations of PE or YE for the determination of maximum growth rates (tmax) or maximum optical density $_{(ODmax).}$ Thus, YE were added at 1 g/L for strains EQ28, EQ12, and R-058 and at 2 g/L for others cultures. Different levels were selected for media containing only PE. A concentration of 2 g/L of PE was chosen for strains EQ12 and R-083 and 3 g/L for other strains. In AS assays involving mixtures of YE and PE, con-centrations determined for YE alone were used.

Fermentations with Lactobacilus Rhamnosus R-011

LBS broth (propietary formulation, Rosell-Lallemand, Montr6al, Canada) was used to evaluate the growth-promoting properties of various YE, PE, or a mix of YE and PE on *L. rhamnosus* R-011. Extracts were added to the LBS broth at a concentration of 5 g/L. The media were pasteur-ized at 90°C for 30 min and the pH was adjusted to 7.0 with 5 M NaOH before inoculation.

Some fermentations were carried out in the same base me-dium rather than in AS assays to evaluate the capacity of AS analyses in predicting data of 2-L fermentations.

Fermentations of *L. rhamnosus* R-011 were carried out at 37°C in 2.5-L New Brunswick BioFlo 3000 U (New Bruns-wick, Edison, U.S.A.). Two litres of medium was inoculated at 1.5% (v/v) from a fresh culture, and pH was measured ev-ery 12 min. OD (600 nm) of fermented media were read when the pH reached 4.3.

Viable counts were also carried out when pH reached 4.3 by plating appropriate dilutions (0.1% sterile peptone) in *Lactobacillus* MRS agar (BDH, Toronto, Canada). Chain lengths were determined by direct microscopic examination after methylene blue staining ac-cording to Packard and Ginn (1985).

No step for breaking chains prior to CFU analyses was included; thus, the total bacterial population was estimated by multiplying the viable count by the average number of cells per chain (Champagne et al. 1995).

Fermentations with Streptococcus Thermophilus R-083

Broth medium #43 (Rosell-Lallemand) was used to evalu-ate the growth-promoting properties of various types of YE and mixtures of YE and PE on *S. thermophilus* R-083. Ex-tracts were added to the media at a concentration of 5 g/L. The pH of the media was adjusted at 5.0 with 85% phospho-ric acid before pasteurization at 90°C for 30 min. The pH was adjusted to 6.8 with 5 M NH_4OH before inoculation.

Fermentations were carried out at 37°C in 1.5-L Biostat M bioreactors (B. Braun Instruments, Munster, Ger-many). One litre of medium was inoculated at 1.0% (v/v) with a fresh culture, and OD (600 nm) was measured every hour.

The pH was maintained at 5.9 with 5 M NH_4OH after the pH of the media reached this value. Viable counts on fi-nal samples taken from fermentation were obtained by plat-ing appropriate dilutions (0.1% sterile peptone) in M17 agar (BDH).

Statistical Analyses

Three independent replicates were carried out for each treatment, and data are the average of these three assays. Statistical analyses were performed using SAS Institute Inc. (1989) software. Analysis of variance was performed using Duncan's test and significance was defined at P <_ 0.05.

Biomass values in AS assays system were transformed into ln values of OD and plotted against time for the expo-nential growth part of the curve. The slope of the linear por-tion of the curve, giving the $_{gmax,}$ was obtained by simple linear regression.

	OD_{max}			Nmax		
Strain	Mean*	SD	SD % of the mean	Mean	SD	SD % of the mean
Lactobacillus casei EQ28	0.53	0.025	4.8	0.1	0.006	5.9
Lactobacillus rhamnosus R-011	0.55	0.038	6.9	0.08	0.007	8.7
Lactobacillus plantarum EQ12	0.68†	0.024	3.5	0.17	0.01	5.7
Pediococcus pentosaceus EQ44	0.27†	0.057	21.7	0.08	0.008	9.3
Streptococcus thermophilus R-083	0.64	0.020	3.1	0.22	0.018	8.0
Lactococcus lactis R-058	0.52†	0.024	4.6	0.13	0.02	15.2

*Mean $_{ODmax}$ obtained with 11 lots of PE.

†Statistical analysis of variance have shown a significant difference between $_{ODmax}$ obtained with the 11 lots.

Results and Discussion

Two aspects of the growth curve were analysed: (1) maxi-mum biomass obtained (using $_{ODmax)}$ and (2) $_{\mu max.}$ We con-sidered that these values present different properties of the growth supplements. Since $_{\mu max}$ occurred at the beginning of the growth curve, the effect on growth is related to the mere presence of essential as well as stimulatory components, and high concentrations are not required at this point. However, for OD_{max}, the concentration of all the essential growth fac-tors is of prime importance, since growth will stop when only one essential factor has been depleted in the medium.

Variability in the Growth-promoting Properties of PE Lots

Lot-to-lot variability in the composition of YE has been demonstrated by Peppler (1982) and confirmed by Potvin et al. (1997). Variability in ingredient composition used in cul-ture medium for the growth of LAB may impart important yield differences. Evaluation of lot-to-lot variability of the PE was thus an important concern to evaluate the potential of this ingredient for large-scale fermentations.

Eleven lots from the same manufacturing process were evaluated for their growth-promoting properties. For the 11 lots, presented in Table 2, the standard deviation values gen-erally represented less than 10% of the average ODmax or $_{\mu max}$ values. Strains reacted differently to the variability in composition of PE lots and, with regard to $_{ODmax,}$ three of the seven cultures (*L. plantarum* EQ12, *P. pentosaceus* EQ44, and *Lactococcus lactis* R-058) were significantly af-fected by lot source (Table 2). However, $_{\mu max}$ data were not significantly affected by lot source. The variable reaction of strains to lot might be related to the fact that LAB have vari-able requirements in growth factors (Desmazeaud and de Roissard 1994; Hugenholtz et al. 1987). These results add to the literature with respect to the significance of lot variabil-ity of media ingredients on growth of LAB and are the first report on PE in this respect.

Poor growth of *L. acidophilus* R-052 was obtained in the base medium containing PE, and the growth rates were diffi-cult to measure. It was our concern that sensitivity to oxygen might affect the growth of this strain in the multiwell plates. Some additional media formulations were then examined.

In the standard base medium containing 3 g/L of PE, ^{OD}max reached 0.087 on average (data not shown). When 0.05% cysteine hydrochloride or 0.1% ascorbic acid were added, the $_{ODmax}$ values were 0.070 and 0.060, respectively. When ascorbic acid and cysteine were both added, growth was better, with $_{ODmax}$ reaching 0.400. Consequently, these two compounds were added to the base medium for the subse-quent experiments with *L. acidophilus* R-052.

Mixtures of PE and YE in the Basal Medium

Growth of bacteria in base medium containing 100% PE was almost always lower than growth in base media contain-ing a mixture of PE and YE or 100% YE, with both ^{OD}max (Table 3) or $_{\mu max}$ data (Table 4). Thus, PE could not be used alone for the satisfactory production of the lactic cultures tested.

One hypothesis for this observation could be the lower content in nitrogen-based compounds in PE. Indeed, YE typically contain between 50 and 71% "protein" (nitro-gen × 6.25; Peppler 1982), which represents, respectively, 7 and 11% nitrogen, while FNI-400 PE contains only 3.97% total nitrogen, of which 2.25% is a-amino nitrogen. The ni-trogen content of YE, which includes amino acids, peptides, and nucleic fractions, has been shown to be a major contrib-utor to the growth-promoting property of YE (Smith et al. 1975; Gaudreau et al. *1999;* Conway et al. 2000). However, mixtures of 50:50 YE-BS:FNI 400 PE gave higher OD lev-els than YE-BS alone for five out of seven strains.

Therefore other compounds than nitrogen contribute to the biological values of PE. Vitamins could be implicated, since PE con-tains vitamins essential for growth of the LAB (Desmazeaud and de Roissard 1994).

The reason for low nitrogen level in PE is the high starch content. The PE contains 32% starch. Starch is not an essential growth factor for lactic cultures. Removal of starch from PE, by ultrafiltration for example, would thus potentially enable a 32% concentration of growth factors and improve the value of PE as a microbiological medium ingredient. Ultrafiltration of YE has been shown to modify their content in nitrogen (Conway et al. 2000), and studies are currently under way to determine if this is also the case with the FNI-400 PE.

For *L. casei* EQ28, *L. rhamnosus* R-011, and *S. thermophilus* R-083, the highest OD values were obtained with a mixture of PE and YE-BS. This suggests that YE and PE complement themselves for these strains with respect to nutrient compounds. Therefore, producers of starter cultures could find an economic interest in substituting YE with PE.

Furthermore, for *S. thermophilus* R-083, mixing PE and YE had a marked benefit on growth rate. The AS data enable the selection of PE and YE mixtures that would benefit biomass yields while reducing costs. Since *S. thermophilus* appears to benefit so much from the PE–YE mixtures, it was a logi-cal candidate for the studies in bioreactors.

Table 5. Composition of the various media used for the 2-L fermentation assays of Lactobacillus rhamnosus R-011 and biomass obtained (OD or CFU/mL) once pH had reached 4.3.

Supplement (g/L)				Biomass		
YE-LA	YE-X803	PE (FNI-400)	*tf* (h)*	OD	CFU/mL	Chain length†
Base medium: LBS						
—	—	—	*22.3a*	*4.3b*	3.6×10^9*b*	1.2*c*
5	—	—	16.0*c*	8.3*a*	7.5×10^9*a*	*1.8ab*
2.5	—	2.5	16.0*c*	9.6*a*	5.9×10^9*ab*	*1.7abc*
—	—	5	19.7*ab*	6.3*b*	3.8×10^9*b*	2.0*ab*
—	2.5	2.5	16.8*bc*	8.7*a*	3.6×10^9*b*	2.0*ab*
Base medium: ASm						
1	—	—	16.0*c*	1.6*c*	2.6×10^8*c*	2.2*a*
0.5	—	0.5	*17.7bc*	*1.3c*	5.6×10^8*c*	*1.8ab*
—	0.5	0.5	19.0*bc*	1.5*c*	3.8×10^8*c*	2.0*ab*

Note: For a given column, means that are followed by the same letter are not significantly different ($P < 0.05$). —, No addition.

*Fermentation time, which represents the time at which the stationary growth phase is reached. †The average number of cells per chain, as determined by microscopic examination.

When mixtures of YE and PE gave the highest $_{ODmax}$ val-ues of all blends, it was always with the YE-BS brand. This suggested that YE-BS was better suited than the other YE brands for the LAB, but closer examination showed that this was not the case. Indeed, 100% YE-X803 gave the highest OD values for *P. pentosaceus* EQ44, while 100% YE-BS was best for *Lactococcus lactis* ssp. *lactis* R-058 and 100% YE-LA was best for *L. acidophilus* R-052 (Table 3). This study thus adds to the literature with respect to the variabil-ity of YE growth-promoting properties and shows that there is a YE–strain interaction in this respect. This points to the usefulness of evaluating the ingredient source when impor-tant quantities of a given strain need to be produced. To pre-vent such source-related variations, Porubcan and Sellars (1979) have suggested to mix the nitrogen sources. The pres-ent study, aimed at evaluating such blends, is in line with this approach. It is noteworthy that in media containing 100% YE, YE-BS gave the highest $_{\mu max}$ for all strains except *S. thermophilus* R-083 (Table 4), which points to the differ-ent pictures that $_{ODmax}$ ^{and} $_{\mu max}$ provide.

Bioreactor Fermentations of Lactobacilus Rhamnosus R-011

One of the goals of this study was to examine if data ob-tained with AS would be in correlation with those obtained in bioreactors. The growth rates were not measured during fermentation in the bioreactors, but the fermentation time does provide a picture. With *L. rhamnosus* R-011, the media that generated the highest biomass levels were also those in which the fermentation was the shortest (Table 5). In LBS medium, the addition YE and PE supplements reduced the fermentation time (*tf;* Table 5). There was good correlation (R^2 = 0.84) between the *tf* values (Table 5) and the corre-sponding $_{\mu max}$ data (Table 4). This suggests that the growth-rate data obtained in AS assays can be helpful to predict the fermentation time.

The optical density (600 nm) and viable counts (CFU/ mL) in samples were measured at the end of the fermentation and the results are shown in Table 5. There is little information of biomass levels in CFU/mL of *L. rhamnosus* in the litera-ture. Arasaratnam et al. (1996) obtained 3.8 × 10^9 CFU/mL in a whey medium supplemented with YE. Viable popula-tions obtained in this study, up to 7.5 × 10^9 CFU/mL, were thus high and show that the PE and YE ingredients were ef-fective in sustaining good growth of lactobacilli.

Bioreactor viable counts and OD obtained with the ex-tracts added in ASm were not significantly different, pre-sumably because of the small differences in the values obtained, the low supplementation level of YE and PE, as well as because of the inherent error (typically between 10 and 30%) of CFU analyses. The relationship between ^{OD}max data obtained in the AS assays and that of the bioreactor in the ASm showed high correlation (R^2 = 0.93; Table 6). This confirms previous data with *L. acidophilus* (Champagne et al. *1999)* and shows that AS is a good tool to predict bio-mass levels at larger scales if the same media and fermenta-tion conditions are used. Biomass levels reached in the AS medium were significantly lower than those in the industrial LBS medium (Table 5). This is probably related to the lower YE and PE quantities added to the AS media.

The viable counts and OD values obtained in the bioreactors with LBS media showed significant differences (Table 5). As was found in the AS assays, OD values in the fermentations were highest when a mixture of YE and PE was used. Indeed, correlations between AS and bioreactor-LBS data for OD were good (R^2 = 0.87; Table 6). Unfortu-nately, AS data seemed less effective in predicting the via-ble counts in LBS-based media (R^2 = 0.37). One reason was found to be variable bacterial chain length (Table 5). When the viable counts were corrected by multiplying CFU/mL values by the average number of cells per chain (evaluated by direct microscopic counts), thus giving the total populations, the correlation with AS data improved.

When the corrected CFU values were themselves trans-formed into their log values, correlations with AS data improved again (Table 6). The question of biomass determi-nation method is an important feature of growth studies. In this study, it was felt that using two methods would provide a more complete picture. Every method has its advantages and disadvantages. OD reflects the total biomass level, but has the disadvantage of not distinguishing between dead and live cells. In fresh cultures such as these, it can be ex-pected that the content of dead cells is low and that OD is a good indicator of live biomass levels. CFU counts provide a better picture of the live biomass level. However, CFU counts are less precise than OD because of the experimen-tal error that accumulates in the successive dilutions. Fur-thermore, the presence of chains introduces variations, which we tried to quantify by carrying out microscopic ex-amination of the cultures. By multiplying the CFU counts (which basically represents the number of chains) by the average number of cells per chain, we had a better picture of the total viable population. It is difficult to say which method is best. However, most teams present the biomass levels with CFU counts to indicate the level of live popula-tion.

Bioreactor Fermentations of Streptococcus Thermophilus R-083

Some assays were conducted with *S. thermophilus R*-083, since this strain was the one most positively affected by the PE during AS assays, and it was an opportunity to verify these data in bioreactor assays. Furthermore, the in-dustrial production of *S. thermophilus* R-083 is carried out under pH control, which was not the case for *L. rhamnosus* R-011. Thus, the bioreactor fermentations differed from the AS assays in the following three aspects: (1) concentrations of PE and (or) YE, (2) growth medium ingredients other than PE or YE, and (3) pH control.

The final biomass values obtained in *S. thermophilus R*-083 fermentations are shown in Table 7. Viable counts (CFU/ mL) obtained in media with all supplements were not judged to be significantly different. $_{ODmax}$ obtained in broth medium #43 with 100% PE was lower than with 100% YE, as predicted by AS data. The 50% substitution of YE-BS with PE (FNI-400) was successful in sustaining good growth because viable counts and OD were not sig-nificantly different than with 100% YE. Therefore, as the AS assays had suggested, partial substitution of YE with PE is worth considering in the industrial production of *S. thermophilus* cultures.

As with *L. rhamnosus* R-011, relationships between ^{OD}max values obtained with AS assays and bioreactor data were examined (Table 6). When AS and fermentation as-says with *L. rhamnosus* R-011 were conducted in the same medium and with the same fermentation parameter (no ex-ternal pH control), the correlation was very high (R^2 = 0.98) between the two sets of OD data. When the industrial medium was used in the fermentation assays, there was a decrease in correlation between the AS and bioreactor OD data. The relationship was still good with the lactobacilli, but was very poor with the streptococci. The industrial me-dia have a much higher content in the growth supplements and many sources of supplementation are often used. Since data from this study show that interactions occur between YE and PE in AS conditions (Tables 3 and 4), it is possible that such a phenomenon also occurs between YE, PE, and the other ingredients of the industrial media.

CONCLUSION

Many growth media contain mixtures of YE and peptones. However, casein, meat, and soy peptones are generally more expensive than YE, and partially replacing YE with such peptones may not be economically advantageous. New sources must thus be examined. Results showed that PE can-not solely replace YE as ingredients in growth media for lac-tic cultures. However, in some instances a partial substitution of YE with PE was beneficial, and this could be economically attractive, since some PE cost 40% less than YE. However, the beneficial effect of YE replacement with PE varies between strains and YE source, and producers of lactic starters must carry out assays to determine if such a strategy is useful for their particular strains.

AS is a powerful tool for the screening of ingredients aimed for use in microbiological media. When the same me-dium used for AS is used for fermentation assays, this study, as well as previous studies (Champagne et al. *1999)*, have shown that good correlations are found between AS-biomass and fermentor-biomass data. However, this study also pointed to limits of the current AS-prediction method with respect to industrial media. Further studies are currently un-der way to determine if the dilution of complete industrial media, as suggested by Potvin et al. (1997), could provide better AS predictions of the growth of the cultures in bioreactors.

4

Inhibition of Bacilli in Industrial Starches by Nisin

The properties of *Bacillus coagulans* and of other bacilli that contaminate paper and paperboard manufacturing processes were investigated under simulated industrial conditions. Nisin (0.05 to 0.125µg ml-') blocked growth of indigenous bacilli that contaminate sizing starches. *B. coagulans* starch isolates, *B. licheniformis, B. amylolique-faciens,* and *B. stearothermophilus* grew at >50°C in industrial starch and produced a-glucosidase and cyclodextrins.

The industrial isolates and reference strains of *B. amyloliquefaciens, B. cereus, B. coagulans, B. flexus, B. licheniformis, B. pumilus, B. sporothermodurans, B. stearothermophilus* and *Alicyclobacillus acidoterres-tris* were inhibited by <0.125 pg of nisin on agar. *B. coagulans* and *B. stearothermophilus* were similarly inhibited by <0.025 pg of nisin ml-' and by 3 pg of the biocide DBNPA ml-' in industrial starch. *B. licheniformis* and *B. amyloliquefaciens* strains were less sensitive. About 40% of nisin added to starch was retained after cooking. Fifty percent of the nisin remained active after 11 h of storage at 60°C. The results show that nisin has potential as a preservative for modified industrial starches. *Journal of Industrial Microbiology& Biotechnology* (2001) 26, 107-114.

INTRODUCTION

Starches are bulk raw materials for the manufacture of food, paper, textiles, pharmaceuticals, and adhesives. The paper industry in North America uses over 2 million tons per year of modified starches for surface sizing, as wet-end additive and in coating. In Finland the annual use of starch is 250 000 tons by the paper industry and 20 000 tons by the food industry. Microbially deteriorated sizing starch impairs the printing quality of office and journal papers and endangers the hygienic quality of food-packaging paper and board. Although enzymatic liquefaction of starch has been studied intensively, the microbiological events leading to hydrolysis and spoilage of industrial starches and their prevention are poorly understood. There are few published papers on the microbial species capable of spoiling starch in a paper machine environment. The preservation strategies using biocides to prevent spoilage are therefore chosen by trial and error.

Biocides are toxic by definition and may be sensitising. Surface - sizing starch is spread on paper or board at the dry end of the machine, representing the immediate interface to consumers' skin or to food in types of paper having no polyethylene extrusion coating. Understanding of microbial starch spoilage and the antimicrobial sensitivities of the spoilage organisms will facilitate the development of consumer-friendly starch-preserva-tion strategies. This paper describes aerobic spoilage microorganisms in industrial starches and their abatement using nisin.

MATERIALS AND METHODS

Bacterial Strains, Inoculation and Media

Aerobic starch-degrading isolates were picked from nutrient agar (LAB M, Bury, England) or from starch nutrient agar (nutrient agar with 0.2 wt.% of soluble starch, Merck, Darmstadt, Germany) incubated at 28 or 50°C. Origins of isolates and reference strains used in this study are listed in Table 1.

The bacterial density of the cultures was measured with a laser nephelometer (Minineph®, The Binding Site, Birmingham, UK) calibrated using Polybead® polystyrene 3 - μm (B. *cereus)* or 1-μm (other bacilli) microspheres (PolySciences, Washington). Indus-trial starch medium contained chemically oxidised surface-sizing potato starch (20 g 1-′), yeast extract (0.1 g 1-′), Na2HPO4 (16 mg 1-′), malic, citric and lactic acids (50 mg 1-′ each) and Tween 80 (0.1 vol.%), pH 6.1. All chemicals used were obtained from local sources and are of analytical quality unless otherwise stated.

Inhibition Assays

The stock solution of commercial nisin (Sigma, St. Louis, MO), 50 μg (2000 IU) active substance ml-′ in 0.02 M HCl with 0.2% Tween 80, was stored at -20°C in the dark and neutralised with one volume of 0.02 M NaOH before dilution with water for use. 2.2 - Dibromo - 3 - nitrilopropionamide (DBNPA) was used as the tradeware Daracide 7819 (Betz-Dearborn, Helsinki, Finland) containing 9% to 12% w/vol of the active substance.

The resazurin reduction assay was performed on a microtiter plate (96 wells) with 330 μl of inoculated (10^6 cells ml-′) industrial starch medium and 20 μl of resazurin (sodium salt, 0.05 mg ml -′ in water, ICN Biomedicals, Ohio) per well supplemented with nisin, DBNPA, EDTA, or nil as indicated. The plate was incubated stationary at 45°C and the fluorescence of the emerging semireduced form ofresazurin (resorufin) was recorded at intervals of 15 min using the kinetic fluorometer Fluoroskan Ascent (Labsystems, Helsinki, Finland) and the filter pair 544/590 x4m.

For the kinetic turbidometric assay a 100-well microtiter plate was filled as above except that no resazurin was used. Optical density was measured at 45°C using Bioscreen C (Labsystems) and a wide-band filter (420-580 nm). The plate was shaken before each measurement, once an hour.

The inocula were grown and the agar diffusion assay was carried out at 37 or at 50°C on brain heart infusion (BHI) agar (Difco Laboratories, Detroit, MI) pH 7.0, except for *A. acidoterrestris*, which was grown on potato dextrose agar (PDA, Difco), pH 3.5. An overnight culture of the test bacterium, 10^6 cells in 3.5 ml semisolid agar with 0.1 vol.% Tween 80, was spread on agar plates (90 mm) with 0.1 vol.% Tween 80. Nisin solution (5 pl) was placed on the inoculated agar surface and the plates were read after 24hat37orat50°C.

Analytical Protocols

Whole-cell fatty acid analysis was performed as described by Vaisanen *et al* except that the biomass was grown at 50°C. Sherlock Microbial Identification System with the aerobic TSBA library version 3.90 (MIDI, Newark, DE) was used to identify the fatty acids. Partial 16S rRNA gene sequences comprising 488 nucleotide positions at the 5'-end of the gene were determined as described previously.

ATP content was measured using the ATP biomass kit (1243 - 118, BioOrbit, Turku, Finland) and a BioOrbit 1253 luminometer. Thermotolerant viable counting was done on plate count agar (PCA, Biokar Diagnostics, Beauvais, France) at 50°C; plates were read after 2 days.

To measure the a-gucosidase activity of the *Bacillus* strains, industrial starch medium was inoculated with 10^6

bacteria ml^{-1} and incubated for 18 h at 37 or at 50°C. a-Glucosidase activity of the cultures was measured using methylumbelliferyl-a-glucoside (MUF; Sigma) as substrate (1.4 mM final concentra-tion). The culture (50 µl) and the MUF-substrate stock (10 µl) were dispensed in each well of the 96-well microtiter plate and incubated at 55°C. After 15 min pH was adjusted to 8.9 by adding 10 µl of 1 M Tris-HCl buffer to each well, and the fluorescence was measured using the filter pair 355/460 run in the Fluoroskan Ascent. For the blank the fluorescence was measured without incubation.

Starch depolymerization was tested on agar containing 15 g soluble potato starch (Sigma S -4251) , 2.5 g NaNO3, 1 g KzHPO4, 0.6 g MgSO4·7H20, 0.1 g CaC12·2H20, 0.1 g NaCl, 1 mg FeC13 and 10 g agar per liter, pH 7. The plates were incubated for 7 days at 37, 50 or 55°C and cooled to +4°C. Depolymerization of starch was visible as a clearing zone around the colony and cyclodextrin accumulation as a zone of white precipitate inside the clearing zone (van der Veen and Dijkhuizen, personal communication).

Nisin activity in industrial starch medium (pH 5.5) was bioassayed as described by Wahlstrom and Saris using L. *lactic* subsp. *lactic* LAC182. Starch was sampled in 5 mM HCl (1:1), stored (<24 h) at room temperature in the dark and analysed for residual raisin. A standard curve was prepared using commercial nisin (Sigma) in industrial starch medium.

RESULTS

Twenty-six thermophilic starch-degrading isolates of *Bacillus coagulans* were isolated from cooked surface-sizing starches, sampled from 10 storage tanks and feed tanks in paper mills. They grew well at 50°C and poorly at 28°C.

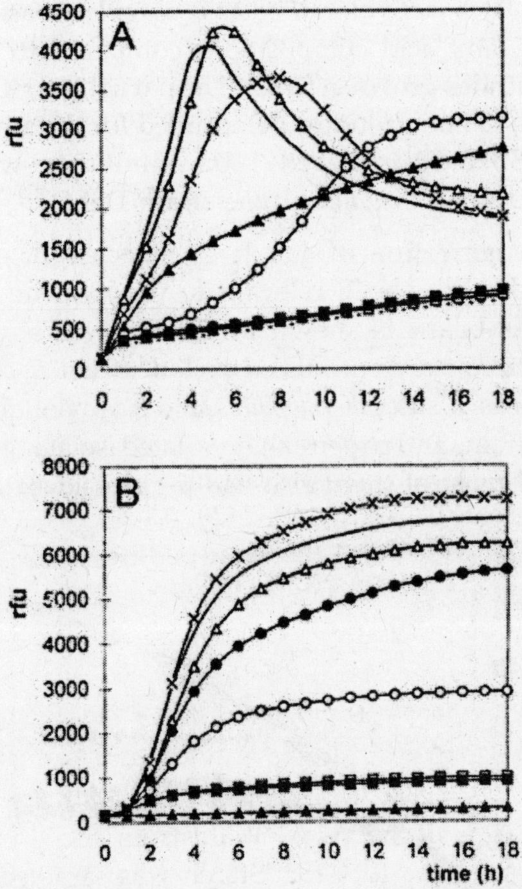

Figure] The metabolic inhibition by nisin, by 2,2-dibromo-3-nitrilopropionamide (DBNPA), and by EDTA of B. coagulans S11 (A) and of B. stearothermophilus DSM1550 (B). Cooked 2% industrial starch medium (pH 6.1) was inoculated with 10^6 cells ml $^{-1}$ and incubated without shaking at 45°C. The curves represent the time course of resazurin reduction as averages of four parallel measurements. Reduction of resazurin into resorufin was recorded by its fluorescence at 590 nm. The fluorescence curves obtained after treatments with 0.125 or with 0.5 µg of nisin ml^{-1} (not shown), with or without EDTA, were similar to those with 0.025 µg of nisin ml -1 (-) No added inhibitor; (o) 0.0025 µg nisin ml^{-1}; (Ï%) 0.0025 µg nisin ml $^{-1}$ + 2 $µg_1$ EDTA ml - ^{1}• (q) 0.025 µg nisin ml $^{-1}$; (n) 0.025 µg nisin ml^{-} + 2 pg EDTA ml^{-} ^{1}; (“) 0.5 µg DBNPA ml^{-} (^{2}%) 3 µg DBNPA ml $^{-1}$; (x) 2 pg EDTA ml $^{-1}$; (....) no inoculum.

The isolates were identified as B. *coagulans* based on the whole-cell fatty acid compositions compared to the type strain (all strains grown at 50°C). The partial 16S rRNA gene sequences (488 nucleotides) determined for the strains S11, S29 and SB2b-50-1 shared 100% similarity with that determined for *B. coagulans* type strain DSM1^T.

Depolymerization of starch by selected *B. coagulans* strains and other bacilli relevant to the hygiene of food-grade paper-board is shown in Table 1. All strains that depolymerized starch produced cyclodextrins. *B. coagulans, B. licheniformis, B. stearothermophilus, an*d B. amyloliquefaciens reference strains and papermaking-related isolates grew well at 50T and most of them also had a-glucosidase activity.

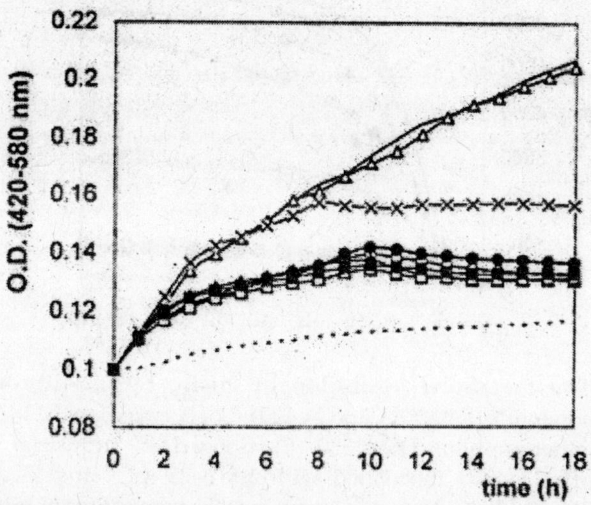

Figure 2 Growth inhibition of starch-degrading *B. coagulans* S11 by nisin, DBNPA and EDTA. The figure shows the time course of optical density increase of microliter plate cultures of the strain S11 in cooked 2% industrial starch (pH 6.1) at 45°C as averages of five parallel wells. Turbidity curves obtained after treatments with 0.125 or with 0.5 pg nisin ml^{-1} (not shown), with and without EDTA, were similar to those with 0.025 μg of nisin ml^{-1}. Symbols are as in Figure 1.

B. licheniformis, and B. amyloliquefaciens were less sensitive. Three industrial contaminant strains of *B. cereus* and the type strain of *B. licheniformis* were only inhibited transiently by nisin.

Metabolic inhibition by raisin and by DBNPA, a commonly used biocide, was studied in thermotolerant starch-degrading strains of *Bacillus.* Figure 1 shows the metabolic sensitivities of an industrial starch contaminant, *B. coagulans* S11 (Figure IA), and of *B. stearothermophilus* DSM1550 (Figure 1B) in industrial starch medium. With no added inhibitor, *B. coagulans* S 11 showed a fluorescence peak, indicating reduction of resazurin, 4.5 h after inoculation with 10^6 cells ml-′, a usual level of contamination in deteriorating starches in the process lines. Three micrograms of DBNPA ml -′ was required to attenuate the metabolism of *B. coagulans* S 11 for > 18 h. The same effect was achieved with 0.0025 µg of nisin ml -′ when 2 µg of EDTA ml-′ was present. In the absence of EDTA, 0.025 µg of raisin ml-′ was needed. EDTA alone (2 µg ml-′) postponed emergence of the fluorescence peak of *B. coagulans* S11 for 3 h. Treatments with >0.025 µg nisin ml-′, with or without EDTA, completely blocked the metabolism of B. *coagulans* S11 and of *B. stearothermophilus* DSM1550.

The sensitivities of the metabolism of *B. coagulans* S 11 (Figure IA) and five other *Bacillus* strains to nisin and DBNPA in industrial starch medium are compiled in Table 3. It shows that 0.0025 to 0.025 µg nisin ml -′ blocked the metabolism of all *B. coagulans* strains during the measurement period of 18 h, whereas *B. licheniformis and* B. amylolique-faciens were less sensitive. The industrial *B. coagulans* isolates were as sensitive as the type strain.

The data in Table 3 and Figure 1 thus show that 0.0025 to 0.025 µg raisin ml-′ were equally or more effective than

3 μg of DBNPA ml⁻′ against all *B. coagulans and* B. stearothermophilus strains. Industrial contaminant strains of *B. licheniformis and* B. amyloliquefaciens were equally sensitive to 0.5 μg of nisin ml⁻′ and to 3 μg of DBNPA ml $^{-1}$. in nutrients and already contained an unknown amount and type of biocide supplied by the starch manufacturer, supported only slow growth (μ<0.5 h^{-1}). The presence of >0.0025 μg nisin ml- ^{1} blocked growth. Results of tests similar to those presented in Figure 2 were obtained also for other industrial contaminant strains and for reference strains of *B. coagulans, B. licheniformis* and *B. amyloliquefaciens,* as summarised in Table 4. B. *stearothermophilus* DSM1550 did not grow measurably under the conditions used.

The areas beneath the growth curves of four *B. coagulans* strains were effectively diminished by 0.0025 to 0.025 μg nisin ml^{-1}, compared to uninhibited controls. B. *licheniformis and* B. amyloliquefaciens strains were inhibited by >0.125 μg nisin ml^{-1}. Inhibition by raisin of the *B. coagulans* type strain but not of the other strains was enhanced by 2 μg EDTA ml^{-1}.

The effects of nisin and DBNPA on the ATP contents and on cultivability of the contaminant bacilli were measured after 18 h incubation. Nisin (0.0025 to 0.025 μg ml $^{-1}$) suppressed viable counts (Table 5) of the tested strains of *B. coagulans* and of *B. stearothermophilus.*

For *B. licheniformis and* B. amyloliquefaciens strains, > 0.5 μg nisin ml $^{-1}$ was required to obtain the same level of inhibition as by 3 μg of DBNPA ml^{-1}. Industrial contaminant B. *coagulans* cultures (strains S11, S29, SB2b-50-1) lost >99% of ATP when 0.025 μg nisin ml^{-1} was present. Paperboard contaminant *B. licheniformis* TSP 29a and *B. amyloliquefaciens* TSP55 lost 88% and 91% of ATP, respectively, when 0.5 μg nisin ml $^{-1}$ was present. Nisin sensitivity of the strains measured by ATP assay thus supported results obtained by the fluorometric resazurin reduction test. This was expected as both methods reflect effects on energy metabolism.

The conclusion from the results presented above is that the starch-degrading thermotolerant bacilli were effectively inhibited by nisin under conditions simulating those prevailing in the paper industry. To see the effect of nisin on a mixed culture situation such as is likely to occur in the industry, cooked surface-sizing starches from storage tanks (80°C) and from feed tanks (46-65°C) were sampled (Table 6). These contained the biocides used by the mills at the time of sampling (information on the effective substances or concentrations not available). Three (A, B, C) of the 10 samples, taken during 4 months, contained 50-400 cfu of thermotolerant (50°C) bacteria ml^{-1}. Nisin (0.05 or 0.125 µg ml^{-1}) effectively suppressed the thermotolerant colony counts of these during 24 h of incubation in the laboratory (Table 6). This is sufficient for industrial application, as the storage time of cooked starches at the mills is usually only a few hours.

Because the biocides used to preserve surface-sizing starches in paper mills gradually lose their activity, they are repeatedly dosed, i.e., into the starch slurry, into the storage tank of the cooked starch, and into the feed tank. To determine the stability of nisin, industrial starch medium containing 0.025 µg nisin ml^{-1} was "cooked" (5 min, 90°C), stored at 60 or 80°C and the residual nisin activity measured. Calibration showed that the accuracy of the assay in industrial starch was optimal at assay dilutions of 0.001 to 0.01 N of nisin ml^{-1} (Figure 3A). The persistence of active nisin in starch during cooking and storage at 60 or 80°C was measured. About 40% of the added nisin (1 N ml^{-1}) remained active after cooking, 50% of that remained active after 11 h of storage at 60°C, and somewhat less after storage at 80°C (Figure 3B). Biocide-treated starches were sampled at the mills. The effective substances were routinely rotated at intervals of 2 to 4 weeks in the mills. Total thermotolerant colony counts (PCA 2 days, 50°C) were analysed in duplicate and the average is shown. The duplicates deviated <0.2 log units. [a]Not tested.

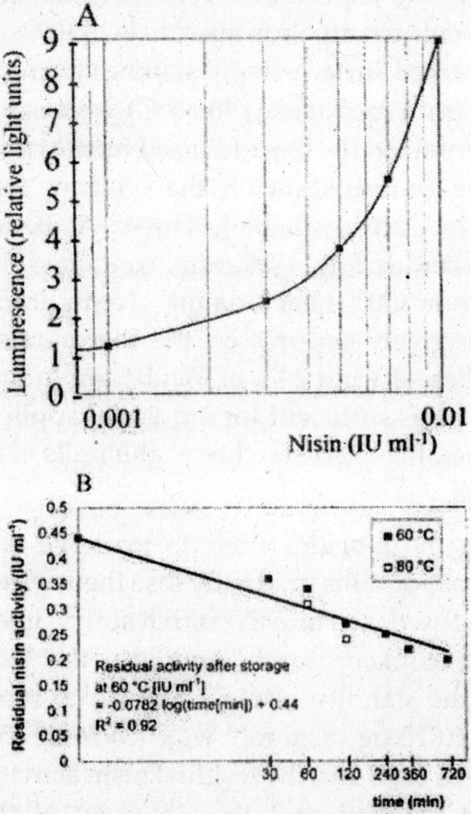

Figure 3 Assay and survival of nisin in cooked industrial starch. (A) Calibration curve for nisin obtained with *L. lactic subsp. lactic* LAC182 in industrial starch medium. The emitted luminescence by the strain LAC182 was measured after 3 h incubation with known amounts of nisin. 1 IU = 0.025 µg of active substance. (B) Residual nisin activity after storage at 60 or 80°C.

DISCUSSION

Bacterial contaminants are not evenly distributed inside paper and paperboard. Confocal laser scanning microscopy analyses have shown that the greatest number of microbes is located just underneath the polyethylene extrusion coating, and likely origi-nated from surface- sizing starch. The microbiological quality of surface- sizing starch is thus important for the hygienic quality of food-packaging paper and board.

In this study we used several parallel methods to test the sensitivity of paper machine and starch contaminants to antimicrobials. The fluorometric resazurin-resorufin method was superior to the other methods used. This method is targeted to detect metabolic activity, indicated by reduction of resazurin. Resazurin becomes fluorescent at the semireduced state, indicat-ing accumulation of NADH inside cells. The fluorometric resazurin-resorufin test measures bacterial metabolic activity rather than growth, and was therefore suitable for food-industry applications. Our results show that it was useful for detecting starch spoilage in the paper industry because it is possible to observe emergence of the fluorescence peak long before any significant increase in turbidity. Compared to plate counting, the fluorometric test was less laborious and shortened the time required for spoilage detection from several days to less than 1 day. ATP determination was about equal in sensitivity but more laborious.

Turbidometry has been used succesfally in inhibition studies of bacilli from paperboard in process waters but its use for monitoring microbiological contamination in starch is seriously limited by the turbidity of most industrial starches. Similar problems were observed with paper industry mineral pigments. The fluorometric resazurin reduction-based assay may also be an option for the microbiological quality control of other highly turbid raw materials like pulps, resin sizers, neutral sizers, and latexes.

The spoilers of starch in the paper industry are aerobic sporeforming bacteria. Bacterial spores in starch survive cooking and storage at 60-80°C and may proliferate in the machine feed tank maintained at 45-65°C. *B. amyloliquefaciens, B. coagulans, B. licheniformis, and* B. stearothermophilus grow at 50-55°C and produce heat-stable starch-hydrolysing enzymes. *B. circulars and* B. subtilis rarely grow at >50°C but do produce heat-activable a-amylases and a-glucosidases with optimum temperatures of 50 to 90°C [9,40,43 1. Formation of cyclodextrins from starch by cyclodextrin glycosyltransferases (CGTases) and their subsequent uptake and intracellular degrada-tion by cyclodextrinases (CDases) is an alternative pathway to the combined actions of amylases and exo-glucosidases, especially among different bacilli.

B. coagulans was the most common contaminant in the surface - sizing starches analysed in this study. Cyclodextrins were produced by all starch depolymerizing isolates from paper industry -related materials, and also by the other tested strains belonging to the species *B. amyloliquefaciens, B. cereus, B. coagulans, B. flexes, B. licheniformis,* or *B. stearothermophilus.* The results indicate that cyclodextrin production is an important pathway for degradation of starch in the paper industry environment. a-Glucosidase activity was also commonly found.

Bacillus species are difficult to eliminate from the industrial process environment using biocides. Nisin (E234) is an antimicrobial peptide produced by *Lactococcus lactic subsp. lactic.* It is authorised for food preservation in more than 50 countries and it has been used to prevent spore outgrowth, e.g., in processed cheese, cheese spreads, canned vegetables and other heat-processed foods. Heating sensitizes spores to nisin.

Nisin in concentrations of 0.0025-2.5 µg ml^{-1} (=0.1-100 N ml^{-1}) is useful for inhibiting different bacilli, this paper]. Spores of thermophilic bacteria such as *Bacillus stearothermophilus and* B. coagulans have been reported to be particularly sensitive [6,13,33 1. Both vegetative cells and spores of bacilli are sensitive to nisin.

Our results show that 0.05-0.125 µg of nisin ml^{-1} (2-5 N ml^{-1}) effectively preserved surface- sizing starches sampled from paperboard processes. Pure cultures of *B. coagulans and* B. stearothermophilus (industrial starch contaminants and reference strains) were even more sensitive. The concentration of nisin required to inhibit *B. licheniformis and* B. amyloliquefaciens in industrial starch (0.5 µg ml - 1) was close to that reported to be inhibitory to *B. licheniformis* spores in milk (25 N ml^{-1}). Our results show that a major amount of nisin remained active under conditions simulating cooking and storage of preserved starch in the industry.

Although nisin inhibited all bacilli tested in the agar diffusion assay, *B. cereus* strains (type strain, industrial, environmental and food poisoning isolates) were among the least sensitive. Maintain-ing high temperature during storage of cooked starches (> 50°C) is thus important for controlling growth of *B. cereus* and other mesophilic bacilli.

A. acidoterrestris and B. sporothermodurans are relatively new food contaminant organisms and have not yet been reported in packaging materials. *B. sporothermodurans* produces exceptionally heat-resistant endospores and the acid-tolerant and heat-resistant *A. acidoterrestris* is a worldwide spoilage problem in pasteurized and heat-treated juices. We show here that these species were also effectively inhibited by raisin, analogous to the findings for *A. acidoterrestris* in fruit juices.

Blackburn observed that the inhibitory effect of nisin against vegetative Gram-positive bacteria at pH >5 was improved by the presence of 0.01-3 mM EDTA (= 3 -1000 μg ml $^{-1}$) . We found synergism with 2 μg EDTA ml^{-1} under paper machine-simulating conditions with one *Bacillus* strain out of seven strains tested. Higher EDTA concentrations were not tested because they are not technically applicable in paper mills.

Our results show that nisin has potential for preservation of starch in industrial processes (Salkinoja-Salonen and Pirttijarvi, patent applied for).

5

Growth Characteristics and Oxidative Capacity of Acetobacter Aceti

We studied the growth characteristics and oxidative capacities of *Acetobacter aceti* IFO 3281 in batch and chemostat cultures. In batch culture, glycerol was the best growth substrate and growth on ethanol occurred only after 6 days delay, although ethanol was rapidly oxidized to acetic acid. In continuous culture, both glycerol and ethanol were good growth substrates with similar characteristics. Resting cells in a bioreactor oxidized ribitol to L-ribulose with a maximal specific rate of 1.2 g g^{-1} h^{-1}).

The oxidation of ribitol was inhibited by ethanol but not by glycerol. Biomass yield (Y_{SX}; C-mmol/C-mmol) on ethanol and glycerol was low (0.21 and 0.17, respectively). In the presence of ribitol the yield was somewhat higher (0.25) with ethanol but lower (0.13) with glycerol, with respectively lower and higher CO_2 production.

In chemostat cultures the oxidation rate of ribitol was unaffected by ethanol or glycerol. Cell-free extract oxidized ethanol very slowly but not ribitol; the oxidative activity was located in the cell membrane fraction. Enzymatic activities of some key metabolic enzymes were determined from steady-state chemostat with ethanol, glycerol, or ethanol/glycerol mixture as a growth limiting substrate.

Based on the measured enzyme activities, metabolic pathways are proposed for ethanol and glycerol metabolism.

INTRODUCTION

Acetic acid bacteria are well known for their ability to convert ethanol to acetic acid. Although the strains in this family have been studied comprehensively and used in industrial processes, only a few metabolic studies are available. According to Nakayama and De Ley (1965), acetic acid bacteria have two enzymatic systems for the conversion of ethanol to acetaldehyde: one is linked to the cytoplasmic membrane and the other is in the cytoplasm.

The membrane-bound alcohol dehydrogenase (ADH), first reported by Nakayama (1961b), was later characterized by Matsuhita et al. (1992) as being a quinohemoprotein–cytochrome c complex located on the periplasmic side of the cytoplasmic membrane. According to Nakayama (1961b), the membrane-bound ADH has more than a hundred times higher specific activity than the soluble enzyme. Likewise, acetaldehyde dehydrogenase is present both as a membrane-bound cofactor-independent enzyme (Nakayama 1961a) and a cofactor-dependent soluble enzyme (Nakayama and De Ley 1965).

Reichstein (Reichstein and Grussner *1934*) was the first to notice bioconversion of ribitol to ribulose by an unidentified acetic acid bacterium. Bhuiyan et al. (*1998*) studied this phenomenon later in the production of L-lyxose from ribitol in a two-step process involving acetic acid bacteria and immobilized L-rhamnose isomerase and D-tagatose 3-epimerase.

Recently, Adachi et al. (*2001*) showed that ribitol oxidation is carried out by two different enzymes; one NAD(P)-dependent cytosolic enzyme and another membrane-bound non-NAD(P)-dependent enzyme. The cytosolic enzyme oxidized ribitol to D-ribulose and the membrane-bound to L-ribulose. Interest in the production of rare L-pentose sugars such as L-xylose and L-ribose is rapidly growing owing to their use in the manufacture of antiviral drugs (Scott *2002*).

We have studied the production of L-ribose and L-arabinose from ribitol which is oxidized to L-ribulose by acetic acid bacteria and converted to L-ribose and L-arabinose by the recently discovered epimerase activity of xylose isomerase (Pastinen et al. *1999*), using cross-linked enzymes in a column reactor (Leisola et al. *2001*; Jokela et al. *2002*). In this study we characterized the growth and ribitol oxidation capacities of *Acetobacter aceti* IFO 3281 in batch and chemostat cultures. We also studied some key enzyme activities in metabolism of glycerol and ethanol and suggest metabolic pathways for these substrates.

MATERIALS AND METHODS

Organisms, Maintenance, and Inocula Preparation

Acetobacter aceti IFO 3281 was stored as a frozen stock culture containing 10% (v/v) glycerol in 2 ml ampoules at –80°C. The inoculum for shake flask cultivations was prepared by cultivating *A. aceti* in a 250 ml Erlenmeyer flask containing 50 ml of the complex medium at 30°C for 20–24 h.

For bioreactor cultivations, and for bioconversion experiments in shake flasks and in a bioreactor, the cells from the preculture were separated by centrifugation (8,000 rpm, 10 min) and then suspended in 50 ml of deionized water to give an initial biomass concentration of 0.1–0.3 g/l cell dry weight (CDW) in the bioreactor. For bioconversion reactions, the cells were suspended in 50 mM sodium phosphate or maleate buffer, pH 7, to give an initial CDW of 2.6–3.0 g/l in shake flasks and 1.7 g/l in the bioreactor.

For bioconversion experiments in a Bioscreen C automatic incubator, the cells were cultivated in a complex medium having 1% (w/v) glycerol as a carbon source. The cells for the experiment were separated (8,000 rpm, 10 min) and resuspended in 50 mM sodium phosphate or maleate buffer, pH 7.

Media

The complex medium contained yeast extract (Difco Laboratories, USA), 10 g/l; trypticase peptone (Becton Dickinson, USA), 10 g/l; KH_2PO_4, 0.125 g/l; $MgSO_4 \cdot 7\ H_2O$, 0.25 g/l; and glycerol (Becton Dickinson) 10 g/l.

The mineral medium was prepared according to Verduyn (*1991*) and contained $NH_4(SO_4)_2$, 5.0 g /l; KH_2PO_4, 3.0 g/l; $MgSO_4 \cdot 7\ H_2O$, 0.5 g/l; EDTA, 15.0 mg/l; $ZnSO_4 \cdot 7\ H_2O$, 4.5 mg/l; $CoCl_2 \cdot 6\ H_2O$, 0.3 g/l; $MnCl_2 \cdot 4\ H_2O$, 1.03 g/l; $CuSO_4 \cdot 5\ H_2O$, 0.3 g/l; $CaCl_2 \cdot 2\ H_2O$, 4.5 g/l; $FeSO_4 \cdot 7\ H_2O$, 3.0 g/l; $NaMoO_4 \cdot 2\ H_2O$, 0.4 g/l; H_3BO_4, 1.0 mg/l; KI, 0.1 mg/l; and silicone antifoaming agent, 0.05 mg/l (BDH Laboratory Supplies, UK). After sterilization, a filter-sterilized (Sterivex-GV, 0.22 m; Millipore, USA) vitamin solution was added, giving a final concentration of biotin 1.05 mg/l, calcium panthotenate 1.0 mg/l, nicotinic acid 1.5 mg/l, myo-inositol 25 mg/l, pyridoxal hydrochloride 1.5 mg/l, *para*-aminobenzoic acid 1.0 mg/l, riboflavin 1.5 mg/l, folic acid 1.0 mg/l, and vitamin B_{12} 1.0 mg/l.

In the chemostat cultivations, three carbon sources and their (1+1) combinations (C-mmol/C-mmol) were investigated. Ethanol was added to the medium without sterilization. Glycerol was heat sterilized and ribitol (Pfanstiehl, USA) was filter-sterilized separately before addition. The pH of the medium was adjusted to 5.8 in all the growth experiments.

Cultivation Conditions

Shake flask cultivations were done in Erlenmeyer flasks (250 ml, 200 rpm) containing 50 ml of the mineral medium at 30°C for 48–52 h. The medium contained 5–10 g/l of the various substrates and their combinations.

Batch and chemostat cultivations were carried out in a 3 l bioreactor (Biostat MD; B. Braun Melsungen, Germany) with a working volume of 1.5 l and automatic pH and temperature control. The cultivation temperature was 30°C and stirrer speed 600 rpm. The culture pH was kept at 5.8 by the addition of 3 M KOH and 0.5 M H_2SO_4. The airflow rate was set to maintain the dissolved oxygen concentration above 30% and it was controlled with a mass flow controller (Bronkhurst HiTech, Ruurlo, The Netherlands).

In the chemostat cultivations, the working volume was kept constant by removing the effluent with a peristaltic pump (Watson-Marlow 101U, UK) connected to a PID-controlled level probe. The dry weight of the cultivation, O_2 consumption, and CO_2 production were measured on successive days to confirm that a steady state had been established.

Bioconversion Experiments

Bioconversion experiments were carried out at 30°C in 50 mM maleate buffer (pH 7). First experiments were done in Bioscreen C (Labsystems Oy, Finland) automatic incubator, which allows 200 samples to be run at a time in the 500-l wells of two honeycomb plates. 350 l of the cell suspension in a buffer solution was added to 50 l of substrate dissolved in the corresponding buffer. The wells were incubated for

3 days with continuous shaking using 100 g/l ribitol, xylitol or L-arabitol as substrate. Bioconversion experiments in shake flasks (200 rpm) were carried out with 50 g/l ribitol for 4 days. Bioconversion experiments in 1-liter bioreactors (Biostat Q; Braun Biotech International, Germany) were carried out with a working volume of 500 ml at 800 rpm. Aeration was set at 1 vvm. The cells were incubated with 50 g/l ribitol for 21 h.

Analytical Methods

Gas Analysis

Carbon dioxide, oxygen, argon, and nitrogen were determined by mass spectrometer (VG-Prima 600, UK). For calculation of the rates of oxygen consumption and carbon dioxide production a temperature of 30°C was assumed and air pressure was taken from the daily weather report.

Optical Density

The optical density of the cell suspension at 600 nm was measured using a spectrophotometer (Shimadzu UV-2100, Japan). The samples were diluted to obtain an optical density in the range 0.1–1.0.

Dry Weight Measurement

Culture samples (10 ml) were vacuum-filtered through a pre-dried and weighed nitrocellulose filter (0.45 m; Schleicher and Schuell, Germany), washed with Milli-Q water, and dried in a microwave oven for 20 min (Ignis, Japan). Two parallel measurements were made on each sample.

Substrate and Metabolite Analysis

Ethanol, glycerol, acetate, succinate, and lactate concentrations were determined by HPLC using an HPX-

87H$^+$ Aminex ion-exclusion column (300×7.8 mm; Bio-Rad, USA) and detected with a Waters (USA) 410 refractive index detector and a Waters 486 UV detector. The column was eluted at 65°C with 5 mM H_2SO_4 at a flow rate of 0.6 ml/min. Ribitol, ribulose, and xylitol were separated on an HPX-87P Aminex ion-exclusion column (300×7.8 mm; BioRad) and detected with a Shimadzu (Japan) refractive index detector. The column was eluted at 85°C with H_2O at a flow rate of 0.6 ml/min. The standard error of the analytical methods was less than 5%, except for ribulose. The ribulose quantification gave results that were too high compared with the theoretical yields. This is due to the lability of ribulose in the standard and in the sample. Therefore ribulose concentrations were estimated based on ribitol consumed. No other sugar metabolites were detected in ribitol bioconversion experiments.

Carbon Balance and Degree of Reduction Balance

An estimation of 50% carbon content of *E. coli* cells (Neidhardt et al. *1990*) was used in calculations of the elemental carbon content of the cell dry weight samples. The carbon balance was calculated as the ratio of carbon recovery in metabolic products, including biomass, to the consumption of substrate carbon. The degree of reduction balance was calculated on the basis of metabolites produced and oxygen consumed.

Preparation of Cell-free Extract

Cells were harvested by centrifugation at +4°C, washed once with 10 mM potassium phosphate buffer (pH 7.5, +4°C) and resuspended in 4 ml sonication buffer (100 mM potassium phosphate, 2 mM $MgCl_2$, 1 mM DDT, pH 7.5). Cells were disrupted by sonication with 0.7 mm glass beads at 0°C in a Vibra Cell sonicator (VCX600, USA). Ten intervals of 30 s pulses were applied with a 30 s cooling period between each pulse. Whole cells and cell debris were removed by centrifugation at +4°C and the clear supernatant, containing typically 2–5 mg protein/ml, was used for enzyme assays.

Protein Content

Protein content of the whole cell samples was determined by a modified biuret method (Verduyn et al. *1990*). The protein concentration in the cell-free extract was determined according to Lowry et al. (*1951*).

Enzyme Assays

These were performed at 30°C with freshly prepared cell-free extracts using a Shimadzu UV-2100 spectrophotometer (Shimadzu, Kyoto, Japan). Reaction rates, corrected for endogenous rates, were proportional to the amounts of extracts added. Glycerol phosphate dehydrogenase (GPDH, EC 1.1.1.8) activity was measured according to Blomberg and Adler (*1989*). Pyruvate kinase (PYK, EC 2.7.1.40), pyruvate carboxylase (PYC, EC 6.4.1.1), pyruvate decarboxylase (PYDC, EC 6.4.1.1), phosphoenolpyruvate (PEP) carboxykinase (PEPCK, EC 4.1.1.32), and malate synthase (MS, EC 4.1.3.2) activities were measured by the methods of de Jong-Gubbels et al. (*1995*). PEP carboxylase (PEPC, EC 4.1.1.31) activity was measured by the modified method of Maeba and Sanwall (*1969*). Phosphoenolpyruvate synthetase (water dikinase pyruvate, PPS, EC 2.7.9.2) activity was measured according to the modified method of Cooper and Kornberg (*1969*).

Alcohol dehydrogenase (ADH, EC 1.1.1.1) and acetaldehyde dehydrogenase (ALD, EC 1.2.1.5 for NAD^+, EC 1.2.1.4 for $NADP^+$) activities were measured according to the methods of Postma et al. (*1989*). Isocitrate dehydrogenase (IDH EC 1.1.1.41) and ribitol dehydrogenase (RDH, EC 1.1.1.56) activities were measured by the methods of Verduyn et al. (*1985*). AcetylCoA synthase (ACS, EC 6.2.1.1) activity was measured by the method of van den Berg et al. (*1996*), citrate synthase (CS, EC 4.1.3.7) by the method of de Jong-Gubbels et al. (*1996*), malate dehydrogenase (MDH, EC 1.1.1.37) by the method of Smith (*1983*), and malic enzyme (ME, EC 1.1.1.39) by the method of Outlaw and Springer (*1983*).

One unit of activity was defined as the amount of enzyme catalyzing the conversion of 1 mol substrate per minute. Specific activities were expressed as milliunits per milligram protein. The enzyme activities of a steady-state sample were measured twice using two different aliquots of cell-free extract samples.

Alcohol dehydrogenase and ribitol dehydrogenase activities in the cytosolic membrane of *A. aceti* were studied by incubating cell membrane fractions discarded from the preparation of cell-free extracts in 5–10 ml of 10 mM maleic acid buffer (pH 7) containing either 70 g/l ethanol or 10 g/l ribitol as a substrate.

RESULTS

Growth Experiments

Preliminary growth experiments on *Acetobacter aceti* IFO 3281 were done in shake flasks on a mineral medium containing ethanol, glycerol, ribitol, xylitol, and their combinations as a carbon source. No growth occurred in a medium containing ribitol or xylitol as the sole source of carbon. Xylitol remained intact in the medium while ribitol was almost fully consumed in 48 h. To our surprise ethanol did not support the growth of *A. aceti* on a mineral medium but was instead converted into acetic acid.

In controlled bioreactor cultivations all the ethanol was converted to acetic acid within 30 h (Fig. 1). Only after 6 days of cultivation did the bacteria start to use the acetic acid in the medium.

The utilization of acetate was accompanied by biomass production detected as a change in optical density of the medium, OD_{600}, and as increasing CDW. The final biomass concentration achieved was 0.63 g CDW/l. The biomass yield, Y_{SX}, on acetate was 0.04 C-mol/C-mol (0.04 g/g) and the maximum growth rate, $_{max}$, was 0.06 h^{-1}.

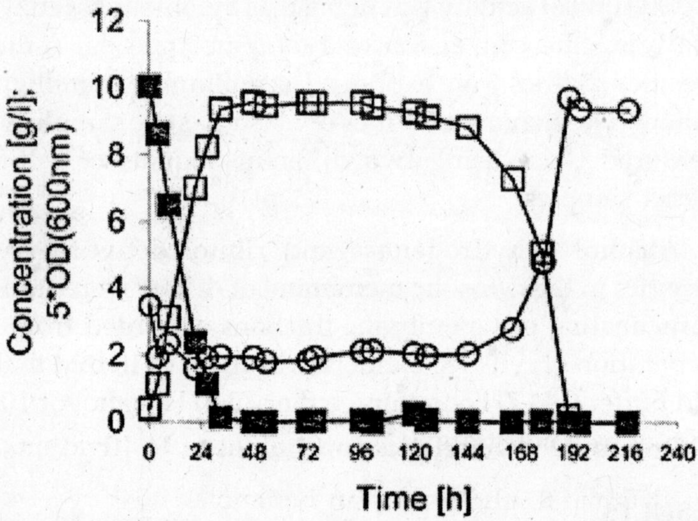

Fig. 1. Acetobacter aceti IFO 3281 cultivated on a mineral medium containing ethanol as the sole carbon source. Ethanol (black squares), acetic acid (white squares) and biomass (white circles) measured as optical density at 600 nm. The values for OD are multiplied by five

When cultivated on glycerol, the growth started slowly and the exponential phase was reached only after 30 h (Fig. 2).

The biomass yield, Y_{SX}, was 0.21 C-mol/C-mol (0.16 g/g) and the maximum growth rate, $_{max}$, was 0.15h^{-1}. The final biomass concentration achieved was 1.26 g(CDW)/l. In addition to biomass, 1.1 g/l lactate and 0.1 g/l succinate were produced. The cultivation was finished after the exponential growth phase had ended and the dissolved oxygen value had returned to full saturation (no oxygen consumption). Under these conditions the residual glycerol concentration was 0.45 g/1.

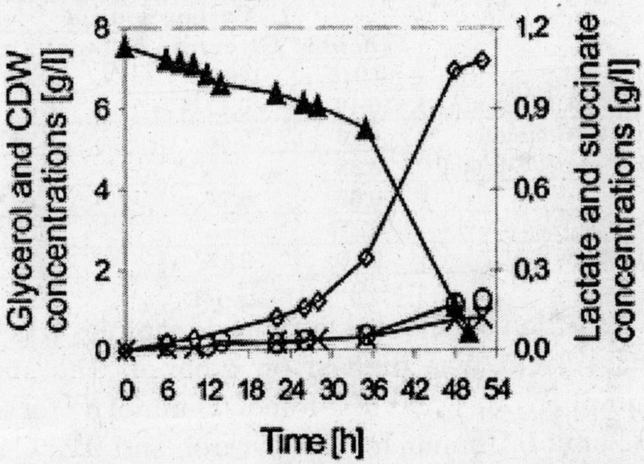

Fig. 2. Acetobacter aceti IFO 3281 cultivated on a mineral medium containing glycerol as the sole carbon source. Glycerol (black triangles), lactic acid (white diamonds), succinic acid (crosses) and biomass (white circles) measured as CDW

Growth Characteristics in Chemostat Cultivations

Ethanol, glycerol or an equimolar mixture of ethanol and glycerol were compared as carbon sources in carbon-limited chemostat cultivations on a mineral medium. The dilution rate was 0.036±0.004 h^{-1}, which is well below the $_{max}$ for both substrates.

The results of the cultivations are presented in Table 1. Biomass and carbon dioxide were the main products of the cultivation, with trace amounts of lactate and succinate accumulating in the medium when glycerol was used as a substrate.

Table 1. Comparison of ethanol and glycerol as a carbon source in a continuous carbon-limited chemostat cultivation of Acetobacter aceti *IFO 3281 on a mineral medium. Biomass yields on substrate, O_2 consumption, CO_2 production, respiration rates, carbon balances, and degree of reduction balances of the cultivation*

	Carbon source		
	Ethanol	Glycerol	Ethanol + glycerol
D [1/h]	0.040	0.034	0.033
Y_{SX} [C-mmol/C-mmol]	0.19	0.17	0.15
O_2 [mmol/C-mmol]	6.16	5.48	6.28
CO_2 [C-mmol/C-mmol]	3.58	4.29	5.01
RQ	0.58	0.78	0.80
C-balance [%]	87.7	90.5	91.6
DR-balance [%]	90.6	91.7	80.6

The carbon balance in steady-state samples was 90±2%. Biomass yield was highest on ethanol. The substrate consumption rate was 0.18 C-mmol/C-mmol h^{-1} for ethanol, 0.24 C-mmol/C-mmol h^{-1} for glycerol, and 0.22 C-mmol/C-mmol h^{-1} for the equimolar mixture of ethanol and glycerol. In the mixture the consumption rates of ethanol and glycerol were 0.10 and 0.12 C-mmol/C-mmol h^{-1}, respectively. The oxygen consumption rate per biomass was essentially the same in all the steady-state experiments and thus independent of the substrate used for growth. The carbon dioxide production rate per biomass, on the other hand, was less when *A. aceti* was cultivated on ethanol than on glycerol or on the mixture of ethanol and glycerol. The respiration quotient was 0.58, 0.78, and 0.80 for ethanol, glycerol, and for the mixture of ethanol and glycerol, respectively. The balances for the degree of reduction were 81–92%.

Enzyme Activities

The metabolic pathways used by *A. aceti* for utilization of ethanol and glycerol were studied by measuring key enzymatic activities in cell-free extracts of the steady-state chemostat samples.

The results of the enzymatic assays are presented in Table 2. Very weak soluble alcohol dehydrogenase activity and an NADP/NADPH-dependent acetaldehyde dehydrogenase activity, more than ten times the ADH activity, were detected in all the cell-free extracts.

Table 2. Enzymatic activities measured in cell-free extracts of steady state samples taken from the chemostat cultivation of* Acetobacter aceti *IFO 3281. The studied carbon sources were ethanol, glycerol and equimolar mixture of ethanol and glycerol

Enzyme	Enzyme activity [mU/mg]		
	Ethanol	Glycerol	Ethanol + Glycerol
Pyruvate decarboxylase	–	2.12±1.12 (NADH)	0.64±0.21 (NADH)
Pyruvate carboxylase	0.73±0.40 (NADH)	0.52±0.04 (NADH)	0.87±0.27 (NADH)
Acetaldehyde dehydrogenase	2.92±0.42 (NADP)	5.00±1.55 (NADP)	1.76±0.17 (NADP)
	1.01±0.13 (NADPH)	1.73±0.16 (NADPH)	0.32±0.02 (NADPH)
Alcohol dehydrogenase	0.16±0.002 (NAD)	0.17±0.001 (NAD)	0.15±0.005 (NAD)
AcCoA synthetase	0.98±0.01 (NAD)	–	0.62±0.50 (NAD)
Isocitrate dehydrogenase	150±90 (NAD)	270±90 (NAD)	Not measured!
	450±130 (NADP)	660±10 (NADP)	
Citrate synthetase	2.95±1.25	2.15±0.83	1.74±0.24
Malate dehydrogenase	–	–	–
Malate synthetase	3.30±0.46 (NAD)	–	1.96±0.07 (NAD)
Malic enzyme	0.65±0.001	–	0.95±0.01
PEP carboxylase	3.56±0.17 (NADH)	5.53±0.17 (NADH)	12.98±0.10 (NADH)
	0.28±0.03 (NADPH)	0.30±0.02 (NADPH)	1.41±0.33 (NADPH)
PEP synthetase	63.36±36.42 (NADH)	39.24±15.97 (NADH)	15.59±3.36 (NADH)

The AcCoA synthase activity was not found when glycerol was the sole source of carbon and energy. On the other hand, pyruvate decarboxylase was detected only in the cells cultivated on glycerol. Pyruvate carboxylase activity was detected in all the samples.

From the TCA cycle enzymes the activities of isocitrate dehydrogenase and citrate synthase activity were detected in all the samples, but malate dehydrogenase activity was not. Glyoxylate cycle enzyme, malate synthase, activity was found in the cells grown on ethanol. The activity varied with the amount of ethanol in the medium. In addition, malic enzyme activity was detected in the cells grown on ethanol but not when glycerol was the sole source of carbon.

Gluconeogenic enzyme, phosphoenolpyruvate (PEP) carboxylase, adjusting the intracellular levels of oxaloacetic acid and phosphoenolpyruvate, and PEP synthetase converting PEP to pyruvate were present in all the steady-state samples. All the enzyme activities detected, except for PEP synthetase and isocitrate dehydrogenase activity, were of the same order of magnitude.

Bioconversion Experiments

Bioconversion experiments in Bioscreen C micro-cultivation apparatus showed that ribitol was converted to ribulose. Arabitol was converted to xylulose. Conversion rate of xylitol to xylulose was very low. Conversion yield of ribulose was higher in the maleate buffer than in the sodium phosphate buffer.

Bioconversion of ribitol by growing cells of *A. aceti* in shake flasks was studied in a mineral medium containing ribitol/ethanol or ribitol/glycerol as carbon sources. When ribitol was combined with ethanol, only very little ribulose accumulated in the medium, but all the ethanol was converted to acetic acid.

No growth was detected. In a medium containing ribitol and glycerol both of the substrates were simultaneously consumed during the cultivation. Biomass and ribulose were the only products accumulating in the medium.

Ribitol conversion in a chemostat was studied with ribitol as a sole carbon source and as an additional substrate in ethanol- and glycerol-containing media. The dilution rate in the cultivations was 0.045±0.002 h^{-1}. Ribitol was not used for growth and cells were washed out during the chemostat cultivation obeying the theoretical first-order washout kinetics.

When ethanol or glycerol was added as a co-substrate together with ribitol in the medium, only part of the ribitol was converted to L-ribulose while the rest remained intact in the medium. Ethanol and glycerol were fully consumed and no other products than biomass and carbon dioxide were detected. When compared with the cultivations without ribitol, a slight increase in the specific oxygen uptake was noticed (Table 3).

The specific carbon dioxide production was approximately the same in the ethanol-containing medium with or without ribitol. The biomass yield was higher when the bacteria were cultivated on a mixture of ethanol and ribitol. In the medium containing ribitol and glycerol, the specific carbon dioxide production was substantially higher and the biomass yield lower compared to the cultivation where no ribitol was present. The respiratory quotient of the cells decreased when ribitol was added to cells growing on ethanol, whereas it increased when ribitol was added into a medium containing glycerol.

The specific conversion activity of ribitol to ribulose was approximately the same regardless of the substrate simultaneously used for growth (Table 3). The specific consumption of ethanol decreased from 4.54 to 4.01 when ribitol was added to the ethanol growing cells, whereas the specific glycerol consumption increased on ribitol addition. As a consequence, the amount of carbon source used per moles of ribitol converted is different for ethanol and glycerol growing cells.

Table 3. Ribitol as a carbon source. Comparison of ethanol and glycerol as an additional source of carbon in a continuous carbon-limited chemostat cultivation of Acetobacter aceti *IFO 3281 on a mineral medium. Biomass yields on substrate, O_2 consumption, CO_2 production, respiration rates and carbon balances of the cultivation*

	Carbon source			
	Ethanol	Ethanol + ribitol	Glycerol	Glycerol + ribitol
D [1/H]	0.045	0.046	0.047	0.044
Y_{SX} [C-mmol/C-mmol]	0.21	0.25	0.17	0.13
O_2 [mmol/C-mmol]	5.88	6.70	5.71	6.27
CO_2 [C-mmol/C-mmol]	3.80	3.67	4.49	5.91
RQ	0.65	0.55	0.78	0.94
C-balance [%]	114.7	102.3	95.4	90,3
Ribitol consumption [C-mmol/C-mmol CDW]	–	2.02	–	2.14
EtOH or glycerol consumption [C-mmol/C-mmol CDW]	4.54	4.01	5.78	7.68
EtOH or glycerol consumption [C-mmol/C-mmol ribitol]	–	2.26	–	3.59

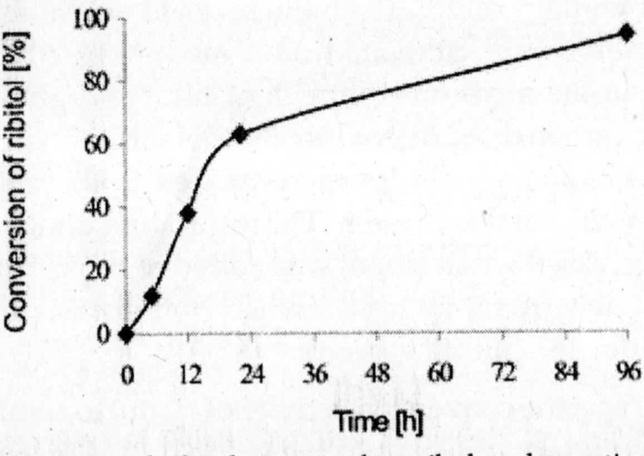

Fig. 3. Percentage of ribitol converted to ribulose by resting cells of *Acetobacter aceti* IFO 3281 in shake flasks containing 50 g/l of ribitol in 50 mM maleate buffer (pH 7)

Resting *A. aceti* cells converted 96% of the 50 g/l ribitol to ribulose with no detectable by-products. The rate of reaction decreased significantly after 24 h when approximately 60% of ribitol was converted (Fig. 3). The activity of the cells was at its maximum 1.0 g g^{-1} h^{-1} at 12 h and at its minimum 0.12 g g^{-1} h^{-1} at 96 h. No oxidation of ethanol or ribitol was detected in cell-free extracts. However, the oxidation activity was in the cell membrane fraction. The rate of ethanol oxidation was 8.7 g l^{-1} h^{-1} and that of ribitol 1.7 l^{-1} h^{-1}.

DISCUSSION

Production of Ribulose

L-Ribulose is an important intermediary compound in the pathway to L-ribose – an important starting material for anti-viral drugs. Production of L-ribulose from ribitol is most conveniently done by resting cells of *Acetobacter aceti* IFO 3281. The reaction first studied by Bhuiyan et al. (*1998*) was adapted and further studied in our experiments. Contrary to the results of Bhuiyan et al. (*1998*), the maleate buffer proved to be more suitable for the reaction than the sodium phosphate buffer. In the bioreactor the cells had the initial activity of 1.20 g L-ribulose formed per 1 g of CDW in an hour, which is almost five times higher than the productivity found in test tubes by Bhuiyan et al. (*1998*).

Ribitol Conversion

The ribitol conversion by *A. aceti* was independent of the growth of the bacterium. Both growing and resting cells were able to convert ribitol to L-ribulose and neither was used in any conditions as a source of carbon or energy. No significant differences in the productivity were detected between growing and resting cells. The dehydrogenase enzyme involved in the conversion is most likely located in the cytoplasmic membrane, since the activity was absent in the cytoplasm, whereas the membrane fraction was able to perform the conversion. Since the reaction has no apparent function in the cellular metabolism, it is assumed that the dehydrogenase enzyme involved is one of the many dehydrogenases found in this strain having an additional activity towards ribitol.

Biomass Yield and Carbon Metabolism

Glycerol was superior as a carbon source for *A. aceti*. The natural substrate for acetic acid bacteria, namely ethanol, served as a substrate for growth in batch conditions only after extended cultivation times. Only after 6 days incubation was the acetic acid, produced from ethanol, used for growth. In continuous cultivation the situation was different: The highest biomass yield on substrate was achieved when ethanol was the sole carbon source. Ethanol, which is favored as a substrate in the natural habitat of this genus, is not immediately used for growth. In batch cultivation the acetate formed from ethanol is consumed only after the energy reserves of the organism have been exhausted. In chemostat, where the cells are forced to grow, ethanol is used as a growth substrate like glycerol. Ribitol and xylitol were not utilized for growth in any of the studied conditions.

Overall, the biomass yield was low compared with the yields reported for other aerobic bacteria (about 0.5 g/g; Nielsen and Villadsen *1994*). A major part of the carbon taken in by the cells of *A. aceti* was metabolized to CO_2. The production of CO_2 was accompanied by a significantly high oxygen consumption rate (17.9–18.9 mmol g^{-1} h^{-1}) compared to the oxygen requirement (8.0 and 10.8 mmol g^{-1} h^{-1}) reported by Brown (*1970*) for *Saccharomyces cerevisiae* and *Escherichia coli*, respectively. The fact that biomass yield in chemostat cultivations on glycerol was smaller than that on ethanol might be due to a more energy-requiring pathway. It is feasible that the uptake of glycerol requires energy, whereas ethanol is taken in by passive diffusion.

Interestingly, when both ethanol and glycerol were present in the medium, the biomass yield was even more reduced in favor of CO_2 production, indicating that the combination of carbon sources was less favorable energetically in providing the precursors for biomass production.

Metabolic Pathways

Very few enzyme activities of acetic acid bacteria have been studied and to our knowledge no systematic studies in continuous cultures on metabolic pathways have been published. Most attention has been given to the so-called vinegar-producing enzymes; alcohol dehydrogenase and aldehyde dehydrogenase. Our results of alcohol dehydrogenase and acetaldehyde dehydrogenase activity measurements in the cytoplasm are in accordance with the findings of Nakayama and De Ley (*1965*). We also tried to detect the activity for glycerol phosphate dehydrogenase in the cell-free extract, but none was found. Since the cells were able to grow on glycerol, they have a pathway for glycerol utilization, and possibly the enzyme is attached to the cytoplasmic membrane.

A. aceti is able to metabolize acetate to carbon dioxide in the reactions of the TCA cycle (Rao *1955*; Stanier et al. *1987*). Isocitrate dehydrogenase and citrate synthase activities were detected in all the steady-state samples, comparable to the activities reported by Weber (*1972*), from cultures containing ethanol as the sole carbon source. Malate dehydrogenase activity, on the other hand, was not found in the cytoplasm. According to Benziman and Palgi (*1971*) the malate dehydrogenase of *A. xylinum* is NAD-independent, and ferricyanide is the most effective co-factor for both the membrane-bound and soluble malate dehydrogenase found in this strain. This may be the case for malate dehydrogenase in *A. aceti* as well.

As a result of active respiration, a large amount of carbon dioxide was produced. The production of CO_2 per biomass was less with ethanol than with glycerol, probably due to utilization of the glyoxylate cycle where no CO_2 is produced. The activity of the glyoxylate cycle enzyme, malate synthase, was dependent on the amount of ethanol in the medium. Additionally malic enzyme activity was detected in ethanol-containing media but not in glycerol media. This is in contradiction to the results of Rao (*1955*), who did not find malic enzyme activity in *A. aceti*.

The AcCoA synthase activity, first demonstrated by Rao (*1955*) and later characterized by O'Sullivan and Ettlinger (*1976*), was not found when glycerol was the sole carbon source. This explains our observation that no acetate was produced in the glycerol medium in the batch experiments. AcCoA for the TCA cycle was either generated from pyruvate by pyruvate dehydrogenase complex or by some other pathway not yet characterized. On the other hand, pyruvate decarboxylase activity found in the cells grown on glycerol would support acetate formation. In either case, phosphoenolpyruvate must first be converted to pyruvate. This reaction was most likely catalyzed by PEP synthetase since pyruvate kinase was absent, as it also was in the studies of Flückiger and Ettlinger (*1977*). In conclusion ethanol was converted to acetate via acetaldehyde followed by conversion of acetate to acetyl coenzyme A. AcCoA then entered the glyoxylate cycle together with oxaloacetate producing energy, reducing power, and precursor metabolites for growth. Oxaloacetate was produced either from malate in the reaction catalyzed by membrane-bound malate dehydrogenase or from pyruvate in the reaction catalyzed by pyruvate carboxylase. Glycerol was most likely metabolized via glycerol-3-phosphate to dihydroxyacetone phosphate and from there on by the reactions of the EMP pathway to phosphoenolpyruvate.

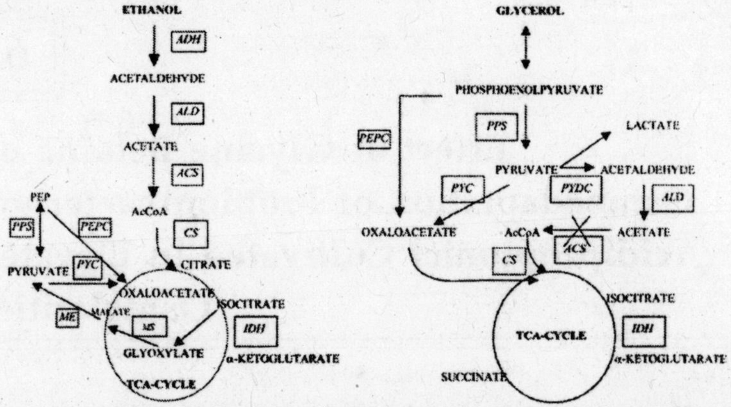

Fig. 4. Schematic presentation of the pathways proposed for utilization of ethanol and glycerol by *Acetobacter aceti* IFO 3821. *Framed boxes* show the enzymes studied. The *crossed arrow* represents an enzyme activity not detected

Although conversion of pyruvate to AcCoA has never been demonstrated in acetic acid bacteria, it was assumed that AcCoA for the TCA cycle was either generated from pyruvate by pyruvate dehydrogenase or more likely via acetaldehyde and acetate in the reactions catalyzed by pyruvate decarboxylase, acetaldehyde dehydrogenase, and membrane-bound AcCoA synthetase. Oxaloacetate was produced from pyruvate by pyruvate carboxylase. Schematic presentation of the proposed pathways is shown in Fig. 4.

Since most of the dehydrogenase enzymes catalyzing the oxidation reactions of interest are located in the cytoplasmic membrane, we are presently characterizing these enzymes. Ribitol to ribulose bioconversion rate increased compared with previous studies. More improvements can probably be achieved by careful optimization of biomass concentrations and additives, and by use of biomass recycle and cell revival operations.

6

Effect of Glycine Betaine on Osmoadaptation of Propionibacterium Acidipropionici Cultivated in Elevated Osmolarities

The sensitivity of industrial strains Acetobacter aceti, Gluconobacter frateurii, and Propionibacterium acidipropionici to osmotic stress was studied. Growth of A. aceti and G. frateurii was totally inhibited at 0.4 M NaCl concentration, but P. acidipropionici was able to grow on a medium containing 1.2 M NaCl.

Addition of glycine betaine to the medium had no detectable osmoprotective effect on A. aceti and G. frateurii cultivations in elevated NaCl concentrations, but it enabled cells of P. acidipropionici to achieve faster the maximum specific growth rate after the prolonged lag phase and therefore to gain faster the final biomass and product concentrations.

The final concentrations of biomass and product of P. acidipropionici were the same as for the cultivations of the bacterium without NaCl and glycine betaine present in the medium. Intracellular accumulation of glycine betaine was detected in P. acidipropionici cells cultivated in the medium containing glycine betaine. The amount accumulated increased with NaCl concentration, suggesting that glycine betaine plays an important role in the osmoadaptation.

INTRODUCTION

The economic feasibility of fermentation processes is often limited by low productivity due to low cell den-sities. In fermentation processes, cells are exposed to osmotic stress caused by medium components and metabolic products. Cells respond to increased osmo-larity by spontaneous water loss, which may inactivate the enzymatic reactions inside the cells and thus be a limiting factor for both growth and production.

Several process alternatives have been proposed to avoid sub-strate and product inhibition, including stepwise sub-strate addition by a fed batch process and continuous withdrawal of the product by a cell recycle process. Although improvements in productivity have been achieved, the cell loss due to osmotic stress has not been avoided.

Bacteria minimize water loss by adjusting their turgor pressure to the external osmotic pressure through the accumulation of inorganic ions and compatible organic solutes. Compatible organic solutes, called osmolytes, when accumulated at high intracellular concentrations are not deleterious to essential biochemical and meta-bolic functions of the cells.

Osmolytes are either syn-thesized de novo (endogenous osmolytes) or taken up from the environment (exogenous osmolytes). Glycine betaine, choline, proline, ectoine, trehalose, and 3-dim-ethylsulfoniopropionate (DMSP) are the most studied among the known exogenous osmolytes (Beumer et al. 1994; Cosquer et al. 1999; Jebbar et al. 1995; Skjerdahl et al. 1996). Endogenous osmolytes synthesized by baç-teria include a few amino and imino acids such as glycine betaine, glutamate, proline, and ectoine; polyol gluco-sylglycerol; and the disaccharides trehalose and sucrose (Gouffi et al. 1999).

Glycine betaine is a compatible solute found in large amounts in some bacterial cells living in environments with high osmolarity. A transport system specific for glycine betaine has been found in both halophilic (Ca-novas et al. 1996) and nonhalophilic bacteria (Bae et al. 1993; Park et al. 1995; Peter et al. 1998; Pichereau et al. 1999; Skjerdal et al. 1995) as well as in methanogenic archae (Proctor et al. 1997). Accumulation of glycine betaine increases cell tolerance to high salinity and drought (Beumer et al. 1994; Jebbar et al. 1995; Kets et al. 1996; Peter et al. 1998; Skjerdal et al. 1996). Be-sides acting as a compatible solute, glycine betaine has an osmoprotective function in some bacteria, stimulat-ing their growth in hyperosmotic environments (Csonka 1989). In addition to glycine betaine transport, *Bacillus subtilis* (Boch et al. 1994) and *Halomonas elongata* (Canovas et al. 1996, 1998) reportedly are able to syn thesize glycine betaine from exogenously added precur-sor choline, and they possess an uptake system specific for choline.

Improvement of the Productivity of Fermentation

Processes requires an enhancement of the osmotolerance of the production strains. It is of utmost importance therefore to understand the mechanism of osmoadap-tation of the bacteria and the osmoprotective effect that the osmolytes have on these bacteria.

The objective of the present study was to establish the effects of glycine betaine addition on bacterial cell cul-tivations in elevated osmolarities created through the addition of NaCl. We examined the osmosensitivity of commercially promising bacterial strains *Acetobacter aceti*,

Gluconobacter frateurii, and *Propionibacterium acidipropionici* and the effect of glycine betaine addition on their osmotolerance. In addition, we studied the uptake of glycine betaine by resting cells of *P. acidi-propionici* with ^{14}C-labeled glycine betaine.

MATERIALS AND METHODS

Organisms, Maintenance, and Inocula Preparation

Propionibacterium acidipropionici ATCC 4875, *Acetobacter aceti* IFO 3281, and *Gluconobacter frateurii* IFO 3254 were stored as frozen stock cultures containing 10% (v/v) glycerol in 2-ml am-poules at -80 °C. Inocula for osmosensitivity studies was prepared by cultivating P. *acidipropionici, A. aceti,* and *G. frateurii* in 250-m1 shake flasks containing 50 ml complex medium at 30 °C for 20-24 *h. A. aceti* and G. *frateurii* were shaken at 200 rpm during cultivation in a rotary shaker (Certomat R, B. Braun Biotech International, Germany).

Inocula for bioreactor cultivation were prepared by cultivating *P. acidipropionici* first in 10-m1 test tubes on complex medium for 24 h at 30 °C and then in 250-m1 shake flasks containing 50 ml complex medium for 24 h without shaking. Cells were separated by centrifugation (8,000 rpm, 10 min, Sorval RC 5 C Plus, Du Pont, USA), washed and suspended in 30 ml deionized water to give an initial biomass concentration of 0.14-0.34 g cell dry weight (CDW) per liter in a bioreactor.

Inocula for glycine betaine uptake studies with resting cells were prepared by cultivating P. *acidipropionici* on 50 ml complex me-dium for 48 h at 30 °C. Cells were separated by centrifugation (8,000 rpm, 10 min, Sorval RC 5 C Plus, Du Pont, USA). Glycine betaine taken up by the cells from the complex medium was washed off with Sorensen buffer solution and washed cells were suspended in deionized water.

Media

Complex medium contained yeast extract (Difco, USA) 10 g/1, trypticase peptone (BBL = Becton Dickinson and Company, USA) 10 g/1, KH2PO4 0.125 9/1, MgSO4·7H2O 0.25 g/1 and glucose (Fluka, Switzerland) or glycerol (BBL = Becton Dickinson and Company, USA) 10 g/1. The pH of the medium was adjusted to 5.8 for *A. aceti* and *G. frateurii* and to 6.8 for P. *acidipropionici* culti-vations.

Mineral medium was prepared according to Verduyn et al. (1992) and contained NH4(SO4)2 5.0 g/1, KH2P04 3.0 g/1, MgSO4 7H2O 0.5 g/1, EDTA 15.0 mg/1, ZnSO4 7H2O 4.5 mg/l, COC12 6H2O 0.3 g/1, MnC12·4H2O 1.03 g/1, CUSO4 5H2O 0.3 g/1, CaC122142O 4.5 g/1, FeSO4·7H2O 3.0 g/1, NaMo042H2O 0.4 g/1, H3BO4 1.0 mg/1, KI 0.1 mg/1, and silicone antifoaming agent

331512 K 0.05 mg/l (BDH = BDH Laboratory Supplies, Eng-land). Glycerol (10 g/1) was added as a carbon source for *A. aceti* and *G. frateurii* cultivations, while glucose (10 g/1) was used in *P. acidipropionici* cultivations.

After sterilization (20 min at 120 °C), a filter-sterilized vitamin solution was added, giving a final concentration of biotin 0.05 mg/1, calcium pantothenate 1.0 mg/1, nicotinic acid 1.0 mg/1, myoinositol 25 mg/1, pyridoxal hydrochlo-ride 1.0 mg/1, and para-aminobenzoic acid 0.2 mg/1.

An additional vitamin supplement was used for acetic acid bacteria cultivation to give final concentrations of biotin 1.05 mg/1, nicotinic acid 1.5 mg/l, pyridoxal hydrochloride 1.5 mg/1, para-aminobenzoic acid 1.0 mg/ 1, riboflavin 1.5

mg/1, folic acid 1.0 mg/1, and vitamin B12 1.0 mg/1. The pH of the medium was adjusted to 5.8 for *A. aceti* and *G. frateurii* and to 6.5 for P. *acidipropionici* cultivations.

Osmosensitivity Studies

Growth of P. *acidipropionici, A. aced,* and *G. frateurii* was studied in Bioscreen C (Bioscreen C, Labsystems, Finland) microcultiva-tion equipment. In Bioscreen C, the changes in optical density in the culture medium due to growth of micro-organism are measured kinetically with a vertical photometer in which the light beam passes up through the bottom of the cuvette and through the sample suspension to a detector. The incubator of the Bioscreen C consists of an incubator tray, a cover for the incubator tray, a temperature control system, and a shaker. Bioscreen C permits 200 samples to be run at a time, with samples placed in the wells of two honeycomb plates, each with 100 wells of 500 0 volume.

In our studies 50 0 inoculum was added to 350 0 mineral me-dium and incubated at 30 °C for 4 days. The optical density of the cultivation medium was measured automatically at 600 nm once an hour. *P. acidipropionici* cultivation plates were shaken 20 s before each optical density measurement and *A. aced* and *G. frateurii* cultivations were continuously shaken.

The NaCI concentration of the medium was adjusted to 0, 0.2_9 0.4_9 0.8, or 1.2 M and glycine betaine concentration to 0 or 20 mM. Glycine betaine (Betafin, anhydrous, pharmaceutical grade) was a kind gift from Finnfeeds, Finland. Results of these experiments are averages of measurements in five to ten different wells. The standard deviation of the optical density measurements of the samples was less than 10%.

Bioreactor Cultivations of P. Acidipropionici

Bioreactor cultivations were carried out in a 1-dm^3 bioreactor (Biostat Q. Braun Biotech International, Germany) with a working volume of 500 ml and automatic pH and temperature control. The temperature of the cultivation was 30 °C and stirrer speed 400 rpm. The culture pH was set to 6.5. The NaCI concentration of the medium was adjusted to 0, 0.4 or 0.8 M. The growth was measured as cell dry weight from the culture medium samples. All experi-ments were done in duplicate. The deviation between the dry weight samples of the parallel cultivation was less than 10%.

Dry Weight Measurement

Culture samples (10 ml) were vacuum-filtered through a predried and weighed nitrocellulose filter (0.45 µm, Schleicher & Schuell, Germany), washed with Milli-p water, and dried in a microwave oven for 20 min (Ignis, Japan). Two parallel measurements were done from one sample and the standard deviation of the technique was determined to be less than 3%.

Substrate and Metabolite Analysis

Samples of 1 ml from the bioreactor cultivations were centrifuged at 10,000 rpm for 10 min (Heraeus Sepatech, Biofuge A. Germany) and the supernatant was stored at -20 °C for further analysis. Glucose, acetate, and propionate concentrations were determined by high-performance liquid chromatography (HPLC). Sample components were separated on an HPX-87H Aminex ion-exclusion column (300 x 7.8 mm, Bio-Rad, USA) and detected with a Waters 410 refractive index detector and a Waters 486 UV detector. The column was eluted at 65 °C with 5 mM H2SO4 at a flow rate of 0.6 ml/min. The standard error of the analytical method was less than 5%.

Intracellular Glycine Betaine Analysis

Culture samples (20 ml) from bioreactor cultivations were centri-fuged (5500 rpm, 15 min, Heraeus Sepatech, Megafuge 1.0, Ger-many) and the biomass was washed once with a fresh mineral medium without carbon source. Washing of the cells was done immediately with a mineral medium of the same osmotic strength as the cultivation medium to avoid the loss of intracellular glycine betaine by osmotic downshock.

Washed cells were suspended in 5 ml of Milli-p water and stored at -20 °C for further analysis. Washed cell suspension (2 ml) was incubated at 100 °C for 10 min and then mixed vigorously with glass beads for 2-10 s in a test tube shaker (Vortex-genie 2, USA). Supernatant was separated by cen-trifugation (10 min, 10,000 rpm, Heraeus Sepatech, Biofuge A, Germany) and glycine betaine concentration was determined by HPLC. The sample components were separated on an HPX-87 C Aminex ion-exclusion column (300 x 7.8 mm, Bio-Rad, USA) and detected with a Shimadzu refractive index detector. The column was eluted at 85 °C with H2O at a flow rate of 0.6 ml/min. The standard error of the analytical method was less than 5%.

Glycine Betaine Uptake Experiments with Resting Cells

The experiment was initiated by adding 3 ml cell suspension to test tubes containing 10 ml mineral medium. To ensure nongrowth conditions the medium contained no carbon source. NaCI con-centration in the medium was adjusted to 0.0 M, 0.2 M, 0.4 M, or 0.8 M and glycine betaine concentration to 0 mM, 1 mM, 5 mM, or 10 mM. Fifty to one hundred microliters of ^{14}C-labeled glycine betaine, prepared by oxidation reaction from ^{14}C-labeled choline, was added to the test tubes to adjust the activity in the medium to 24.3-

31.5 nCi/ml (54,000-70,000 cpm/ml). Labeled glycine betaine was kindly provided by Danisco-Cultor Kantvik Research Center, Finland.

Medium samples of 1 ml were taken immediately after the ad-dition of cell suspension and after 45 and 90 min of incubation at 30 °C without shaking. Samples were filtered through a cellulose nitrate membrane (0.45 µm, Sartorius, Germany) and washed with a mineral medium in which the glycine betaine was replaced with 0.5 M sucrose. Sucrose was added to the washing medium to pre-vent the loss of intracellular glycine betaine during washing by osmotic downshock. Membrane filters were analyzed for /i-radia-tion by liquid scintillation counting (LS 6000 IC, Beckman, USA). The number of counts per minute in the sample was proportional to the amount of glycine betaine taken in by the cells in the sample. All uptake experiments were done in duplicate. The deviation be-tween the liquid scintillation counts of parallel samples was less than 10%.

Results

Osmosensitivity Studies

A. aced (IFO 3281) and *G. frateurii* (IFO 3254) were very sensitive to elevated osmolarities. *A. aced* did not grow when the NaCI concentration of the medium was 0.2 M or higher. *G. frateurii* could grow on medium containing 0.2 M NaCl, but not when the NaCI con-centration was 0.4 M or higher. Glycine betaine addi-tion had no effect on the growth of these bacteria in elevated osmolarities.

As can be seen in Fig. 1, the lag phase of the growth of *P. acidipropionici* increased and the maximum growth rate decreased with increasing NaCI concentration. *P. acidipropionici* (ATCC 4875) was able to grow on a medium containing 1.2 M NaCl, although the maximum growth rate was significantly lower (0.009 absorbance units (AU h^{-1})) than that in a medium without NaCI (0.040 AU $h^{-'}$). Achievement of the maximum growth rate after the lag phase was faster in cultivations where glycine betaine was present in the medium.

No growth was detected for any of the three bacteria on mineral medium containing glycine betaine as the only carbon source, verifying that glycine betaine could not be used as a sole carbon source for growth by these strains.

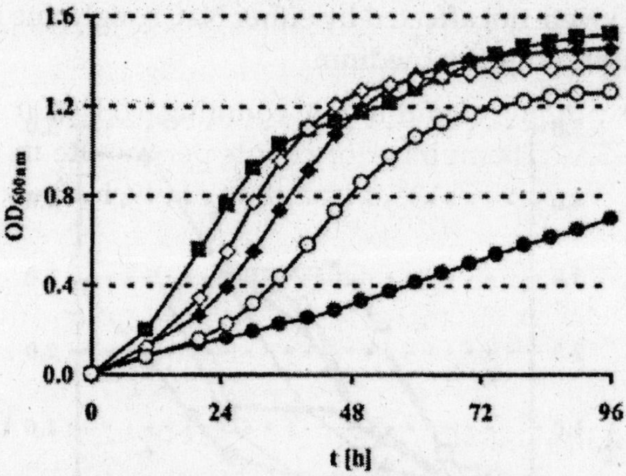

Fig. 1 Effect of increasing osmolarity and glycine betaine on the growth of P. *acidipropionici* (ATCC 4875). Growth was measured as change in optical density of the cultivation medium at 600 nm. The NaCI concentration in the medium was adjusted to 0.0 M NaCI (n), 0.8 M NaCI (¨), or 1.2 M NaCI (n). *Closed symbols* are for cultivations without glycine betaine and *open symbols* for cultivations with 20 mM glycine betaine

Bioreactor Cultivations of P. Acidipropionici

P. acidipropionici (ATCC 4875) adapted faster to ele-vated NaCI concentration when glycine betaine was present in the medium. This result, obtained in the small-scale cultivations, was confirmed in bioreactor cultivations with automatic pH control. The lag phase of the *P. acidipropionici* cultivation increased to 36 h in a medium containing 0.8 M NaCI compared to the 12 h in medium without NaCI (Fig. 2). Although the presence of glycine betaine in the medium had no effect on the lag phase of the cultivation, it did inhibit the effect of ele-vated NaCI concentration on the specific growth rate.

The effect of glycine betaine was seen as faster achieve-ment of the maximum specific growth rate after the lag phase relative to cells cultivated without glycine betaine. Specific biomass yield (YSx) and propionic acid yield (YXp) (Table 1) were not affected by either NaCI or glycine betaine concentration in the medium.

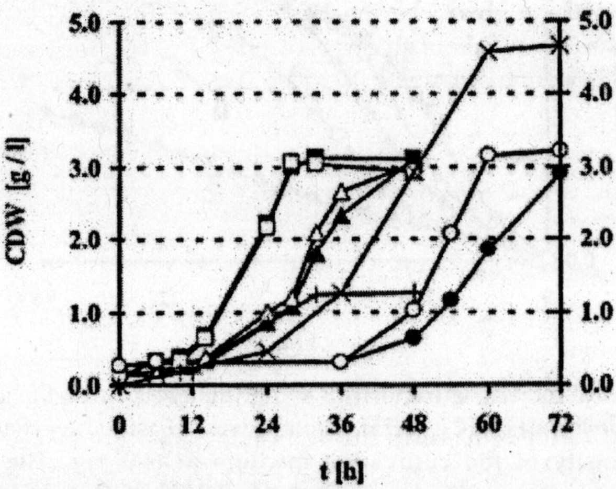

Fig. 2 Effect of increasing osmolarity and glycine betaine on the growth and glycine betaine uptake of P. *acidipropionici* (ATCC 4875) in bioreactor cultivations. The growth was measured as cell dry weight (CDW) on a mineral medium containing 0.0 M NaCI (n), 0.4 M NaCI (A), or 0.8 M NaCI (0). *Closed symbols* are for cultivations without glycine betaine and *open symbols* for cultivations with 20 mM glycine betaine. Intracellular glycine betaine concentration was measured as the difference between the original glycine betaine concentration in the medium and the concentration at the time of measurement in a mineral medium containing 0.0 M (+) or 0.8 M NaCI (x) and 20 mM glycine betaine

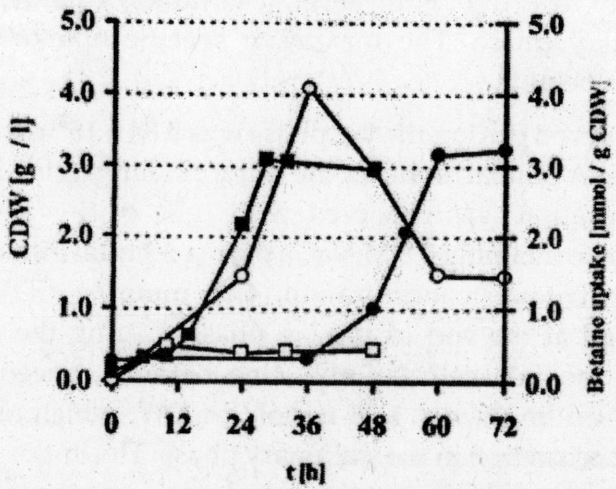

Fig.3 The uptake of glycine betaine by P. *acidipropionici* cell*s* cultivated in a bioreactor. The uptake of glycine betaine is calculated per gram CDW on a mineral medium containing 0.0 M NaCI (q) or 0.8 M NaCI (O). The growth measured as CDW in the same media is represented by *closed symbols*

The uptake of glycine betaine was determined di-rectly by measuring the intracellular glycine betaine concentration of the cells and indirectly as the differ-ence between the initial concentration of glycine betaine in the medium and the glycine betaine concentration at the time of measurement (Fig. 2). During the cultivation, glycine betaine was taken up by *P. acidipropionici* until the stationary phase was reached. Glycine betaine was present in excess in the medium (_13 mM) and was at no time exhausted, suggesting that the uptake was not limited by glycine betaine concentration. As can be seen in Fig. 3, the cells cultivated in a medium without NaCI achieved the maximum level of intracellular glycine betaine concentration, 0.50 mmol/g CDW, at the end of the lag phase.

During the growth phase the intracellular glycine betaine concentration decreased to 0.37 mmol/g CDW, but increased again to near maximum level, 0.43 mmol/g CDW, in the stationary phase. The maximum specific uptake rate of glycine betaine

achieved during the lag phase was 0.84 x 10^{-6} mol/g per minute. A similar trend in the intracellular glycine betaine concentration was observed with cells culti- vated in a medium containing 0.8 M NaCl: the max- imum intracellular glycine betaine concentration, 4.08 mmol/g CDW, was achieved at the end of the lag phase. During the growth phase the intracellular gly- cine betaine concentration decreased linearly to 1.46 mmol/g CDW, which also was the concentration in the stationary phase. The initial specific uptake rate of glycine betaine achieved during the lag phase was 1.00 x 10^{-6} mol/g per minute and comparable to the uptake rate in the medium without NaCl. The maxi- mum specific uptake rate of glycine betaine, 3.74 x 10^{-6} mol/g per minute, was achieved at the end of the lag phase.

Table 1 Maximum specific growth rates, biomass yield on substrate, and propionic acid yield on biomass on mineral medium containing 0, 0.4, and 0.8 M NaCI with and without 20 mM glycine betaine addition. The yields are calculated at the point of cultivation where the concentration of* P. *acidipropionici (ATCC 4875) biomass was 2 g/1

Cultivation conditions		Maximum specific growth rate		Yields calculated at X = 2 g/1	
NaCI	Glycine betaine	μmax	Cultivation	Y_{sX}(biomass/glucose)	Y_{XP}(propionate/biomass)
(M)	(MM)	(per hour)	time[a] (h)	(g/g)	(g/g)
0	0	0.12	19	0.83	0.83
0	20	0.12	19	0.87	0.87
0.4	0	0.13	30	0.74	0.74
0.4	20	0.14	30	0.82	0.82
0.8	0	0.10	51	0.75	0.75
0.8	20	0.12	51	0.68	0.68

[a] Cultivation time at which the maximum specific growth rate was achieved

Glycine Betaine Uptake by Resting Cells

Resting cells of *P. acidipropionici* (ATCC 4875) were able to take in glycine betaine from the incubation me-dium. As can be seen in Fig. 4, the amount of glycine betaine taken up by the resting cells increased with the amount of NaCI in the medium. Even in the medium containing no NaCI some glycine betaine was taken up by the resting cells when these were added to the incu-bation medium after washing of the cells with buffer solution and suspension in deionized water.

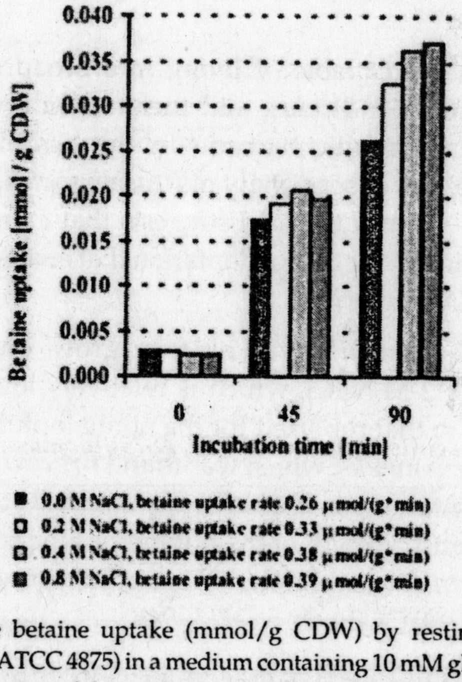

Fig. 4 Glycine betaine uptake (mmol/g CDW) by resting cells of *P. acidipropionici* (ATCC 4875) in a medium containing 10 mM glycine betaine. NaCI concentration of the medium was adjusted to 0.0 M *(black bars)*, 0.2 M *(white bars)*, 0.4 M *(light gray bars)*, or 0.8 M *(dark gray bars)*. The glycine betaine uptake rate was linear in the studied range, regression coefficient $0.980 < R^2 > 0.999$

During 90 min incubation the uptake rate of glycine betaine per cell dry weight was constant and higher in the cells in-cubated under osmotic stress than in the cells incubated in a medium containing no NaCl. The uptake rate of glycine betaine was not directly proportional to the NaCI concentration in the medium since a plateau in the uptake rate (values shown in Fig. 4) was achieved in the medium containing 0.4 M NaCl. Uptake of glycine betaine by the resting cells was independent of the gly-cine betaine concentration in the range of 1-10 mM in the medium, indicating that the uptake of glycine betaine was dependent only on the osmolarity created through the addition of NaCl.

Discussion

Bacteria are capable of living in a broad spectrum of environmental conditions, and there is wide variety even among nonhalophilic bacteria in the osmolarities they tolerate. The lower sensitivity of *P. acidipropionici* than of *A. aced* and *G. frateurii* to NaCI suggests that *P. acidipropionici* has a mechanism for osmoadaptation that enables its growth in elevated osmolarities.

P. acidi-propionici was able to grow on a medium containing 1.2 M NaCl, which is relatively low osmolarity com-pared to that reported for the nonhalophilic bacterium *Staphylococcus aureus*, which was found to grow on a defined medium containing 2.5 M NaCI (Vijaranakul et al. 1997). On the other hand, growth of the nonhal-ophilic soil bacterium *Bacillus subtilis* was inhibited in a medium containing more than 0.6 M NaCI (Boch et al. 1994).

The effect of glycine betaine on growth characteristics of *P. acidipropionici* cultivations in elevated osmolarities was less dramatic than expected. According to Vi-jaranakul et al. (1997), osmoprotectants glycine betaine and choline significantly shorten the lag phase of the NaCl-sensitive *S. aureus* mutants. This was also ex-pected in *P. acidipropionici* cultivations, but no such effect was seen. The results we have presented show that *P. acidipropionici* cells in elevated osmolarities achieved maximum specific growth rate after the prolonged lag phase faster when glycine betaine was present in the medium.

Other growth characteristics were not affected by either NaCI or glycine betaine in the medium. The maximum levels of biomass and product concentrations obtained in the stationary phase were ultimately the same in media with and without glycine betaine addi-tion, although the stationary level was achieved faster when glycine betaine was present. Beumer et al. (1994) found, on the other hand, that the final amount of col-ony-forming units (CFU) of *Listeria monocytogenes* per milliliter cultivated in a medium containing 3 % (0.52 M) NaCI was ten times as great when glycine betaine (1 mM) was present as when it was not. The CFU of the cultivation were not measured in our experiments. Interestingly, two different rates of uptake of glycine betaine were detected during the lag phase of *P. acidi-propionici* cultivation in the medium containing 0.8 M NaCl, suggesting that some cellular reactions are needed for full function of an active transport of glycine betaine. As the prolonged lag phase could not be overcome by exogenously added glycine betaine, it can be presumed that other cellular reactions besides glycine betaine up-take are needed for total adaptation of the cells.

The time needed to achieve the maximum growth rate in the medium may be essential for the synthesis of compatible solutes not present in the medium. Experiments with ^{14}C-labeled glycine betaine and resting cells of *P. acidi-propionici* confirmed that the uptake of exogenously added glycine betaine was not a growth-dependent phenomenon. Glycine betaine was not metabolized and the amount of glycine betaine taken in by the cells, as also the uptake rate, increased with the osmolarity in the incubation medium. We therefore assume that the transport was osmotically induced and glycine betaine acted as a nonmetabolizable compatible solute in adjusting the osmolarity inside the cell to the osmolarity in the environment, in order to maintain the turgor pressure. Glycine betaine has been found to have similar osmotically induced transport in a number of organisms, including *Enterococcus faecalis, Corynebacterium glu-tamicum, Bacillus subtilis, Brevibacterium linens,* and *Yersinia enterocolitica* (Jebbar et al. 1995; Kappes et al. 1996; Park et al. 1995; Peter et al. 1998; Pichereau et al. 1999).

The maximum specific uptake rate of glycine betaine was much lower in the resting than the growing cells of *P. acidipropionici.* Possibly the transport of glycine be-taine requires some energy-dependent cellular reactions. Osmotically induced glycine betaine transport mecha-nism in *P. acidipropionici* was again suggested by the immediate ejection of intracellular glycine betaine from the cells when these were exposed to osmotic downshock by washing with water (results not shown here). A small amount of glycine betaine was also found in cells incu-bated without osmotic stress by NaCl, perhaps in response to the osmotic stress on the cells by other components of the medium. This is consistent with the results obtained for the cultivations of *P. acidipropionici* in elevated osmolarities due to glucose, where no NaCI was added.

The lag phase of the growth was extended with increased glucose concentration of the medium (results not shown here). On the other hand, it may well be that betaine has other functions in the cell besides acting as an osmolyte. Boch et al. (1994) assumed that the choline uptake and conversion to betaine found in *B. subtilis* on a low-osmolarity medium was due to the maintenance of the very high turgor in *B. subtilis*.

In conclusion, exogenously added glycine betaine enabled *P. acidipropionici* to achieve faster its maximum specific growth rate after the prolonged lag phase in elevated osmolarities and so to achieve faster its final biomass and product concentrations. The final concen-trations of biomass and product were the same as when NaCI and glycine betaine were not present in the me-dium. A major improvement in fermentation processes might thus be achieved through the elevated substrate and product concentrations allowed by bacteria when their osmotolerance is enhanced by the addition of gly-cine betaine to the medium.

D-Mannitol (here: mannitol) is a naturally occurring sugar alcohol with six carbon atoms. It is only half as sweet as sucrose. However, mannitol and other sugar alcohols exhibit reduced caloric values compared to the respective value of most sugars, which make them applicable as sweeteners in so-called "light" foods. Moreover, sugar alcohols are metabolized independently of insulin and are thus also applicable in diabetic food products. Besides applications in the food industry, mannitol is also used in the pharmaceutical industry. In medicine, mannitol is used to decrease cellular edema (excessive accumulation of fluid) and increases the urinary output.

In this doctoral thesis, the development of a new bioprocess for the production of mannitol is described. For this purpose, aspects such as strain selection, choice of process method, optimization of process parameters, scale-up, and metabolic engineering were studied. At present, mannitol is produced commercially by catalytic hydrogenation of fructose-containing syrups. The existing chemical production methods are, however, characterized by several drawbacks. The uppermost being that when fructose is catalytically hydrogenated only about 50% of it is converted into mannitol, whereas the rest is converted into another sugar alcohol, sorbitol.

In addition, ultra-pure (expensive) raw materials (fructose and hydrogen gas) are required for efficient conversion. When more cost-effective raw materials, such as glucose-fructose syrups are used as starting material for catalytic hydrogenation, the main product is sorbitol and mannitol is formed as a by-product. Hence, mannitol production becomes very dependent on the market demand of sorbitol. Furthermore, mannitol is relatively difficult to purify from sorbitol. In addition, ion exchange is required for removal of the metal catalyst from the production solution. This results in even higher production costs and decreased yields. The microbial mannitol production process described in this thesis is based on high cell density cultures of slowly growing heterofermentative lactic acid bacteria. The bioconversion of fructose to mannitol was performed in a slowly agitated membrane cell-recycle bioreactor equipped with pH and temperature control.

Neither aeration nor nitrogen flushing of the bioconversion medium was required, which drastically lowers the investment costs of such a plant. An important detail in the new bioprocess was the re-use of cell biomass in successive bioconversions. In a semicontinuous production experiment, the initial cell biomass provided stable mannitol productivities and yields for at least 14 successive batches. Moreover, using a simple purification protocol comprising cooling crystallization of a supersaturated solution and crystal recovery by means of drum centrifugation, high yields of high-purity (>98%) mannitol crystals were obtained. Moreover, in scale-up trials the microbial mannitol production process was successfully run at a small pilot-scale (100 L).

The yield of crystalline mannitol from the initial sugar consumed in the bioprocess was about 52% (w/w). This compares favorably to a commercial chemical process with a yield of about 39%. Hence, under optimized conditions the best production strain (Leuconostoc mesenteroides) converted up to 95% (mol/mol) of fructose consumed into mannitol. Unfortunately, this is only achieved when a significant amount of glucose is co-metabolized by the cells. The catabolism of glucose enables cofactor regeneration in the cells and is thus, essential for the bioconversion of fructose to mannitol.

Moreover, some mannitol is also lost in the purification steps. Another significant improvement brought about by the new bioprocess was a reduced by-product burden. In the commercial chemical process, a total of 1.58 kg by-products are formed for each kilogram of mannitol crystals produced. In the bioprocess, only 0.67 kg by-products are produced per kilogram crystalline mannitol. Using tools of genetic engineering, two key enzymes involved in the primary metabolism of another efficient mannitol-producer, Lactobacillus fermentum, were inactivated. A mutant deficient in D-lactate dehydrogenase and grown in a fructoseglucose medium produced high levels of both mannitol and pure L-lactate. Inactivation of both lactate dehydrogenases resulted in major rerouting of glucose catabolism, which led to the accumulation of pyruvate and production of 2,3- butanediol. Moreover, mutants with lowered fructokinase activity and deficient in acetate kinase were constructed and studied.

CONCLUSIONS

Mannitol has a steady market, but the methods for producing mannitol are still evaluated and developed. Commercial production of mannitol relies on catalytic hydrogenation of fructose or fructose-glucose mixtures. However, the hydrogenation process has several drawbacks. In the hydrogenation process, fructose is reacting with a metal catalyst under high pressure and temperature. The hydrogen needed is supplied in the form of hydrogen gas. With the best metal catalysts available, only about 50% of fructose is reduced to mannitol, while the rest is reduced to sorbitol.

Metal catalysts are non-specific in regards to the substrates, i.e. they catalyze a variety of reactions depending on the substrates present in the reactor. Besides that the catalyst must be of high quality, also the fructose and hydrogen gas must be of high purity to avoid further yield losses and unwanted side-reactions.

A complicated purification process adds additional costs to the catalytic hydrogenation process. First, ion exchange must be used to remove the metal catalyst. Second, high temperature reactions result in color formation and extra purification steps are needed to remove the color impurities. Third, although a big solubility difference between mannitol and sorbitol favors efficient separation of these two compounds, production of high purity mannitol (or sorbitol) requires additional purification steps. Moreover, pure fructose is seldom used as a starting material in commercial production. Instead cheaper fructose-glucose mixtures are used, in which case mannitol becomes the side product of the process. It is thus obvious that the current chemical mannitol production methods are both laborious, ineffective and relatively expensive.

Alternative production methods based on both enzymatic and microbial techniques have been studied. Enzymatic processes for mannitol production often applies socalled cofactor regeneration systems, but the use of these systems on a commercial level, is restricted by factors such as strong end product inhibition, high Km value of key enzyme for fructose and low volumetric productivities. A noteworthy microbial production process was developed by Ojamo et al. (2000). High yields and volumetric productivities were achieved with this bioprocess alternative. However, some drawbacks remained, like cell leakage from the reactor column and slow fructose consumption at low concentrations.

In this work, a well-known enzymatic reaction, in which heterofermentative LAB reduce fructose into mannitol, was applied to develop a commercially competitive mannitol bioprocess. Several LAB species were compared in their ability to produce mannitol and based on the resulting data, an efficient strain belonging to the L. mesenteroides species was identified. When this strain was grown in a simple batch process good yields but only moderate productivities were achieved. To increase the productivity, more sophisticated bioprocess alternatives were studied. Using membrane cell-recycle bioreactor techniques and optimizing the critical process parameters, productivities over 20 g/L/h were achieved. In Table 16, the new bioprocess is compared to the traditional catalytic hydrogenation process.

Regardless of whether the yield or productivity carry more weight from an economical point of view, both parameters ought to be met by the new bioprocess. It is naturally important to remember that a supply of glucose is also needed in the bioprocess and hence, the maximum theoretical yield for the bioreaction phase is only about 67%

of mannitol from the sugar consumed (the real yield for the new bioprocess was about 61-62%).

The functionality of the new bioprocess was tested on two levels. First, on a laboratory- scale, the process was run in a semi-continuous mode a total of 14 batches. Hence, it was shown that stable yields and productivities were possible to achieve in successive batches using the same initial biomass. Second, moving from a 2-L laboratory-scale to a small pilot-scale (100 L), no changes in essential process parameters were observed. It is thus assumed that the new bioprocess will pass further scale-up phases and is applicable for commercial production-scale. Moreover, the scalability of this process concept is expected to be better than e.g. the process using immobilized cells.

The capital costs for the new bioprocess are low compared to the capital costs of typical bioprocesses. Due to a low contamination risk not much effort must be put into maintaining aseptic conditions and only moderately equipped reactors are needed. Also, no gassing systems are required and only the basic parameters (temperature and pH) need to be controlled.

Moreover, the purification comprises basic well-established unit operations. The production costs of the new mannitol bioprocess, on the other hand, are strongly dependent on the price of fructose. One key benefit of microbial processes, in general, is that low-purity raw materials can be used without affecting the yields or productivities. In the case of mannitol production, this will drastically reduce the production costs and is a highly relevant factor, when the new bioprocess alternative is evaluated against the chemical hydrogenation processes. Furthermore, the use of L. mesenteroides, especially in the food industry, is commonly accepted.

The use of LAB as hosts for production of recombinant proteins will most likely grow in importance in the future as new cloning tools are developed. Significant progress has been made with homofermentative LAB, where e.g. Lc. lactis has become a very popular cloning host. Cloning of heterofermentative LAB is also a topic of current research efforts (e.g. Bourel et al., 2001). In order to find applications for the glucose catabolism of the new mannitol bioprocess, metabolic engineering techniques were applied. For instance, mutants producing mannitol and either optically pure L-lactate or pyruvate were constructed. Hence, the total yield of valuable products from sugar consumed was significantly improved in both cases.

In the future, efforts should be placed in finding sources of readily available lowpurity fructose syrups. Ideally this syrup would contain both fructose and glucose (e.g. side streams from a glucose to fructose isomerization process). The effects of using a low purity raw material source must then be carefully validated. After the sugars were depleted from the medium, the cells were observed to slowly consume mannitol. Hence, correct timing of the product recovery by filtration is vital. However, without an on-line sugar analysis system, the correct timing was difficult to estimate. Although some indications were obtained from the base consumption rate, in the future, other on-line analysis methods must be considered and studied. Finally, the reaction rates by which fructose was reduced to mannitol were surprisingly high. Hence, it would be sensible to study the use of this reduction power for reduction of other substrates to their respective high-value reduced forms.

7

Production of Mannitol by Heterofermentative Lactic Acid Bacteria

Eight heterofermentative lactic acid bacteria were compared as to their ability to convert -fructose to -mannitol. Four promising strains were identified and the effects of growth temperature, pH, and nitrogen flushing on mannitol production by these strains were studied in batch bioreactor cultivations. In contrast to earlier findings, a high growth temperature was observed to improve the yield of mannitol from fructose with *Lactobacillus fermentum*. When *Lb. fermentum* was grown at 25, 30, and 35 °C, yields of 86.4±0.8, 88.9±2.4, and 93.6±0.6 mol%, respectively, were achieved. In general, constant nitrogen gas flushing of the growth media was found to improve the mannitol yields, but not the volumetric mannitol productivities. Applying the most promising strain in a batch mannitol production experiment, high average and maximum volumetric mannitol productivities (7.6 and 16.0 g/l/h, respectively) were achieved.

INTRODUCTION

-Mannitol is a six-carbon sugar alcohol, which is about half as sweet as sucrose. It is found in small quantities in most fruits and vegetables [1 and 2] and has various applications, e.g. in foods, pharmaceuticals, medicine, and chemistry. At present, mannitol is produced commercially by catalytic hydrogenation of fructose syrups or invert sugar with the co-production of another sugar alcohol, sorbitol. Typically, the hydrogenation of a 50/50–fructose/glucose mixture results in a 30/70 mixture of mannitol and sorbitol. Besides the fact that mannitol is the by-product of the chemical process and thus can be liable to supply problems, it is also relatively difficult to separate from sorbitol.

In contrast to most sugars and other sugar alcohols mannitol dissolves poorly in water (13% (w/w) at 14 °C) . Cooling crystallization is therefore commonly used to separate mannitol from sorbitol and other components. However, according to Takemura et al. the yield of crystalline mannitol, in the chemical process, is still only approximately 17% (w/w) based on the initial sugar substrates.

In order to improve the total process yield of mannitol it would be advantageous to develop a process with mannitol as the main product and with no sorbitol formation.

Some alternative processes based on the use of microbes have been suggested in the literature. Yeast, fungi, and lactic acid bacteria (LAB) especially, have proved to produce mannitol effectively without co-formation of sorbitol. Among LAB only heterofermentative species are known to convert fructose into mannitol [3, 7 and 8]. Species belonging to the genera *Leuconostoc, Oenococcus,* and *Lactobacillus* particularly, have been reported to produce mannitol effectively. In addition to mannitol these microbes co-produce mainly lactic and acetic acid, carbon dioxide, and ethanol. These by-products are, however, easily separated from mannitol.

Several groups have studied the bioconversion of fructose into mannitol with growing cells, but only two groups report volumetric mannitol productivities over 5 g/l/h. Using a fed-batch cultivation protocol Soetaert et al. reached a volumetric productivity of about 6.3 g/l/h with *L. pseudomesenteroides [9 and 10]. A similar productivity level (6.4 g/l/h) was earlier also achieved with an unidentified Lactobacillus species, named B001 . Moreover, these groups report mannitol yields (mannitol produced from fructose consumed) of 94 and 96 mol%, respectively. Recently, Korakli et al. reported a 100% yield with a strain of Lb. sanfranciscensis . However, this strain grew very slowly and the volumetric productivity in a fed-batch cultivation was only 0.5 g/l/h. Other species of heterofermentative LAB reported to be producers of mannitol include L. mesenteroides, O. oeni, Lb. brevis, Lb. buchneri, and Lb. fermentum [3, 12, 13 and 14].*

Although several papers are available reporting mannitol production with LAB, only one focuses on the comparison of two different species. Yun and Kim cultivated two food-isolated LAB strains in shake flasks, and under optimal growth conditions they concluded that the more effective strain (*Lactobacillus* sp.) converted 86 mol% of fructose consumed into mannitol, whereas the other strain (*Leuconostoc* sp.) had a yield of only 65 mol% . They also reported that from the variety of carbohydrate substrates tested, notable mannitol formation was detected only when either fructose or sucrose were used as the substrate.

In this paper, the production of mannitol by eight heterofermentative LAB was studied. A simplified production (SP) medium was developed and in batch cultivations the effects of temperature, pH, and nitrogen gas flushing on growth and mannitol production were examined. Results reported here enable a preliminary comparison of the behaviour of the studied microbes in bioreactor environments. Moreover, this study identifies two LAB strains with clearly better volumetric mannitol productivities in batch mode compared to prior studies and the strains used in them.

MATERIALS AND METHODS

Microorganisms and Medium

The microbial strains used in this study, summarized in Table 1, were initially grown in a modified MRS growth medium containing 20 g/l fructose and 10 g/l glucose (pH 6.2). *O. oeni* grew poorly at pH 6.2 and was therefore cultured at an initial pH of 5.0. A SP medium was prepared from individual stock solutions. The following concentrations were used in the complete SP medium: tryptone (Pronadisa, Hispanlab S.A., Spain), 10 g/l; yeast extract (Difco, Becton Dickinson and Company, USA), 5 g/l; K_2HPO_4, 2 g/l; fructose, 20 g/l; glucose, 10 g/l; $MgSO_4$, 0.2 g/l; $FeCl_3$, 10 mg/l; $CaCl_2$, 20 mg/l; $MnSO_4$, 10 mg/l, and NaCl, 10 mg/l. The amounts of the variable components (Mg, Fe, Ca, Mn, and Na) were adapted from Dols et al.

Table 2. The simplified production (SP) medium used in the bioreactor cultivations (amounts in brackets were used in Lb. fermentum cultivations)

Component	Concentration (g/l)
Tryptone	10
Yeast extract	5
K_2HPO_4	2
Fructose	20
Glucose	10
$MgSO_4$	0.2 (0.4)
$MnSO_4$	0.01 (0.02)

The final SP medium, used in 600-ml bioreactor cultivations is presented in Table 2. In the 2-l *Lb. fermentum* bioreactor cultivation, SP medium was used with modifications. The initial fructose, glucose, and yeast extract concentrations were increased to 100, 50, and 10 g/l, respectively.

Growth Conditions

The initial comparison of strains and the development of the SP medium were conducted in a Bioscreen C analyzer (Labsystems Oy, Finland). A working volume of 400 l was applied. The temperature was controlled at 30 °C and the optical density (600 nm) of the cell suspensions was measured automatically at 1-h interval. Samples for analysis of mannitol concentration were taken at an early stationary growth phase. Bioscreen experiments were carried out in quadruplicate, but due to the large number of possible samples a cell-free culture broth from four parallel cultures were combined and analysed by high performance liquid chromatography (HPLC).

Bioreactor cultivations were carried out in either a Biostat Q system (B. Braun Biotech International, Germany) with four 600-ml (working volume) culture vessels or a Biostat MD system with a 2-l (working volume) culture vessel. Both systems were equipped with automatic probes for the measurement of temperature, pH, and dissolved oxygen tension (DOT). In the Biostat Q system, magnetic bars and a magnetic drive unit were used for mixing.

The agitation speed was 400 rpm. In the Biostat MD system, two disc impellers were used for mixing (200 rpm). During anaerobic experiments, the growth media were constantly flushed with nitrogen gas, whereas during semi-anaerobic experiments no gases were added to the bioreactors. The DOT probes were calibrated with nitrogen gas and air. The pH was controlled automatically by addition of 5 M NaOH or 2 M H_2SO_4. The 2-l *Lb. fermentum* production study was performed semi-anaerobically at 40 °C and pH 5.0.

All pre-cultures were grown in standard MRS (pH 6.2) (Pronadisa, Hispanlab, S.A., Spain). A 5% (v/v) inoculum of a 10-h culture (*L. mesenteroides, L. pseudomesenteroides, Lb. brevis,* and *Lb. fermentum*), a 20-h culture (*Lb.* sp. B001, *Lb. sanfranciscensis, Lb. buchneri*) or a 3-day-old culture (*O. oeni*) was used. All conditions were examined in two successive cultivations. These results are thus presented as mean values and standard deviations. The 2-l cultivation was inoculated 10% (v/v) with a 10-h culture.

Analytical Methods

Cell growth was monitored by measuring the optical density at 600 nm (OD_{600}) against water. The cell-free extracts were prepared as follows: cells at late exponential growth phase were harvested, washed twice in 50 mM phosphate buffer (pH 6.5) and resuspended in 4 ml of the same buffer. The suspension was sonicated in the presence of glass beads for 8×15 s with 30 s cooling on ice between the treatments.

Cell debris was separated by centrifugation and supernatants were assayed immediately for NADH oxidase activity. NADH oxidase activity was assayed at 30 °C in 50 mM sodium phosphate buffer (pH 6.5) by following the oxidation of 0.13 mM NADH at 340 nm. Protein concentrations were measured by the Bradford method using the Bio-Rad Protein Assay (Bio-Rad Laboratories, USA).

Glucose, fructose, ethanol, and mannitol concentrations were measured with HPLC using an Aminex HPX-87C column (Bio-Rad Laboratories, USA) at 60 °C with Milli-Q water as the eluent. The flow rate of the eluent was controlled at 0.6 ml/min. The HPLC apparatus consisted of a Perkin-

Elmer Series 200 Autosampler and LC Pump (Perkin–Elmer Corporation, USA), and a HP 1047A refractometer (Hewlett–Packard Company, Japan). Also a Deashing Micro-Guard pre-column (Bio-Rad Laboratories, USA) was used. Organic acids were analysed with an Aminex HPX-87H column as described earlier .

The maximum specific growth rates ($_{max}$) were calculated with Microsoft Excel. A chart for the natural logarithm of optical density values versus time was drawn. The maximum specific growth rate is the steepest slope of a linear trendline (3–5 successive values) in the exponential growth phase.

RESULTS AND DISCUSSION

The presentation of the results is divided into three parts. In the first part, experiments conducted in l-scale are presented, whereas the second part covers the results from the 600-ml bioreactor studies. In the third part, knowledge gathered in parts one and two was applied to perform a 2-l mannitol production experiment.

Bioscreen

Comparison of Different Mannitol Producing LAB Species

The results of culturing the strains in the modified MRS medium are presented in Table 3. *L. mesenteroides* ($_{max}$=0.57/h), *L. pseudomesenteroides* (0.46/h), *Lb. brevis* (0.45/h), and *Lb. fermentum* (0.55/h) grew significantly faster than the other four species and as expected, they were also superior in volumetric mannitol productivity.

With most strains tested a significant fraction of fructose consumed by the cells escaped probably into the phosphoketolase pathway and thus, into formation of excess lactic and acetic acid, ethanol, and carbon dioxide. However, in unison with Korakli et al. we found that *Lb. sanfranciscensis* converted almost 100% fructose consumed into mannitol. Based on the productivities shown in Table 3, *L. mesenteroides*, *L. pseudomesenteroides*, *Lb. brevis*, and *Lb. fermentum* were chosen for the bioreactor studies.

Table 3. Mannitol production by heterofermentative LAB

Strain	*r* (g/l/h)	*q* (g/l/h)	*Y* (mole/mole)
L. mesenteroides ATCC-9135	1.53	0.96	0.91
Lb. brevis ATCC-8287	1.47	0.72	0.85
Lb. fermentum NRRL-B-1932	1.45	0.74	0.79
L. pseudomesenteroides ATCC-12291	1.44	0.87	0.86
Lb. sp. B001 BP-3158	0.86	0.44	0.90
Lb. sanfranciscensis E-93491	0.78	0.44	0.98
Lb. buchneri TKK-1051	0.76	0.37	0.84
O. oeni E-97762	0.11	0.12	0.93

Column headings: *r*, the volumetric mannitol productivity; *q*, the specific mannitol productivity (here the volumetric productivity divided by the optical density); *Y*, the yield of mannitol produced from fructose consumed in the initial comparison experiments.

Development of a Simplified Production (SP) Medium

In order to simplify the growth medium for production studies on a bioreactor scale a Bioscreen analyzer and the SP medium were used.

The variable metal components (Mg, Fe, Ca, Mn, and Na) were omitted one at a time from a basic SP medium (tryptone, yeast extract, K_2HPO_4, and sugars). The results showed that only minor changes in maximum specific growth rates were seen, when the variable components were missing from the basic medium (data not shown). The volumetric mannitol productivities, however, were significantly affected by the removal of Mn^{2+} from the growth media of all four strains.

In comparison to the respective values in the complete SP medium, the volumetric productivities of *Lb. brevis, Lb. fermentum, L. mesenteroides* and *L. pseudomesenteroides* in medium without Mn^{2+} decreased 6, 32, 9, and 17%, respectively.

To a lesser degree also the removal of Mg^{2+} was found to decrease the volumetric mannitol productivities (2–4%). Furthermore, a significant decrease in volumetric mannitol productivity was only seen with *Lb. brevis*, when using the simple SP medium instead of the nutrient-rich MRS medium (32%).

In an additional experiment the basic SP medium was supplemented with variable concentrations of Mg^{2+} and Mn^{2+}. The results indicated that the volumetric mannitol productivity was improved even further when double amounts of both Mg^{2+} and Mn^{2+} were used in cultures of *Lb. fermentum*. Table 2 summarizes the composition of the SP medium used in the 600-ml bioreactor cultivations.

Manganese and magnesium ions are essential co-factors for enzymes in the primary sugar metabolism of LAB. Magnesium functions as a co-factor for fructokinase, phosphoketolase, and acetate kinase, whereas manganese functions as a co-factor for some enzymes in the pathway from glyceraldehyde-3-P to pyruvate and for lactate dehydrogenase.

Clearly these metal ions play a central role in the production of reducing power (NAD(P)H) and ATP, and hence, are essential for many cellular functions and more importantly, for the transport and reduction of fructose.

Mannitol dehydrogenase, on the other hand, does not require any metal ions.

Bioreactor Cultivations

Effect of Temperature and pH on Growth and Mannitol Production

The maximum specific growth rates of all the four strains were clearly improved when the growth temperature was increased from 25 to 35 °C (Fig. 1). The growth temperature was also observed to have a strong influence on the volumetric mannitol productivity of the cells. The effect of growth temperature on productivity was particularly evident with *Lb. fermentum*, where a change from 25 to 35 °C brought about an approximately twofold increase in the volumetric mannitol productivity (100±0.02 to 2.03±0.04 g/l/h). Of the four strains tested, *L. mesenteroides* was least affected by changes in the growth temperature.

In fact, increasing the growth temperature from 30 to 35 °C with *L. mesenteroides* resulted in a small decrease in volumetric productivity (1.97–1.94 g/l/h). The specific mannitol productivities (volumetric productivity divided by the optical density) were also clearly higher at 35 °C than at 25 °C. In cultivations with *Lb. brevis*, *Lb. fermentum*, *L. mesenteroides*, and *L. pseudomesenteroides* the specific mannitol productivities were improved from 0.18 to 0.22, 0.16 to 0.30, 0.37 to 0.50, and 0.34 to 0.45 g/l/h, respectively, when grown at 35 °C instead of 25 °C.

On the other hand, better mannitol yields were achieved with *Lb. brevis*, *L. mesenteroides*, and *L. pseudomesenteroides*, when the growth temperatures were lowered (Fig. 1). However, entirely opposite findings were made with *Lb. fermentum*, where a high growth temperature resulted in the best yield. When the temperature was controlled at 25 °C, the yield with *Lb. fermentum* was measured to be 86.4±0.8 mol%. Respectively, at 35 °C *Lb. fermentum* converted up to 93.6±0.6 mol% of the fructose consumed into mannitol.

The highest maximum specific growth rates were achieved when the pH was controlled at 5.5 (Fig. 2). Lower pH values (5.0 and 4.5) decreased the maximum specific growth rates of all four strains.

The maximum specific growth rate of *L. mesenteroides* was especially affected by a low pH. The maximum specific growth rate of this strain, at pH 4.5, was only approximately half of its respective value at pH 5.5. A high pH value improved volumetric mannitol productivities, whereas better mannitol yields were observed at low pH values. The effect of pH on the specific mannitol productivities was found to be small (data not shown).

Earlier Soetaert observed in studies done with *L. pseudomesenteroides* that decreasing the growth temperature and the pH result in more efficient conversion of fructose into mannitol, i.e. a better yield. On the other hand, he also observed that a low temperature and a low pH lead to decreased volumetric mannitol productivities. Similar observations are reported here. For example, in cultivations with *Lb. fermentum* a change in growth temperature from 35 to 25 °C resulted in a 50% decrease in volumetric mannitol productivity.

However, the study also revealed that this behaviour cannot be generalized to all heterofermentative LAB and all temperature ranges. With *L. mesenteroides* a change from 35 to 30 °C resulted in a small improvement of volumetric mannitol productivity. Even more divergent to earlier findings is the observation made with *Lb. fermentum*, where high temperatures resulted, in addition to increased productivities, also in better mannitol yields.

Effect of Nitrogen Gas Flushing on Growth and Mannitol Production

The growth of all four strains was considerably more rapid under semi-anaerobic conditions (i.e. no gassing of the growth media) than under strict anaerobic conditions (i.e. constant nitrogen flushing of the growth media). When compared in maximum specific growth rates, *Lb. brevis* was found to be least affected (up approximately 6%) by the change of anaerobic to semi-anaerobic conditions, whereas the maximum specific growth rates of the other three strains were increased in the range of 15–19%.

The volumetric mannitol productivity of *Lb. fermentum* increased from 1.33±0.02 to 1.65±0.06 g/l/h, when the cells were grown under semi-anaerobic conditions rather than under anaerobic conditions. A more subtle increase was obtained with the other strains, where the volumetric mannitol productivities of *Lb. brevis, L. mesenteroides*, and *L. pseudomesenteroides* were improved with 0.9, 2.9, and 3.7%, respectively. The differences in specific mannitol productivities observed under anaerobic and semi-anaerobic conditions were very small (data not shown). On the other hand, the yield of mannitol from fructose was higher under anaerobic conditions with three of the strains (*Lb. brevis, Lb. fermentum*, and *L. mesenteroides*). The respective behaviour of *L. pseudomesenteroides* did, however, deviate from the former pattern. Under anaerobic conditions the mannitol yield of *L. pseudomesenteroides* was 73.8±0.1 mol%, whereas under semi-anaerobic conditions it was up to 77.1±1.7 mol%.

Hence, nitrogen gas flushing of the growth media seems to be ineffective as a way to assure high volumetric or specific mannitol productivities. In fact, clear enhancement in volumetric productivities was seen with all four strains, when grown under semi-anaerobic conditions rather than under strict anaerobic conditions. The most significant change was again seen with *Lb. fermentum*, where an almost 25% increase in volumetric mannitol productivity was obtained under semi-anaerobic conditions. This observation is naturally at least partly a direct correlation with the improved growth rate of this strain under semi-anaerobic conditions compared to the respective rate under anaerobic conditions. Although the yield of mannitol from fructose was higher in cultivations with constant nitrogen gas flushing (exception: *L. pseudomesenteroides*), it would be more cost-effective to build and run an industrial-scale process without the need to invest in expensive bioreactor gassing systems.

Surprisingly, during the semi-anaerobic experiments oxygen depletion in the growth medium differed notably among the strains. At the beginning of these experiments the DOT in the growth media was approximately 90±5%. During the experiments *L. mesenteroides* was observed to run out of oxygen (DOT=0%) after 2.0±0.1 h, whereas *Lb. brevis*, *Lb. fermentum*, and *L. pseudomesenteroides* ran out of oxygen at 5.4±0.5, 6.7±1.0, and 7.0±0.4 h, respectively. The NADH oxidase activities at t=7 h in the semi-anaerobic experiments were as follows: *Lb. brevis*, 0.57 U/mg protein, *Lb. fermentum*, 0.20 U/mg, and *L. mesenteroides*, 0.49 U/mg. No activity was detected with *L. pseudomesenteroides*. Hence, the specific activity measured for *L. mesenteroides* is similar to earlier reports, 0.6 and 0.44 U/mg [18 and 19].

The lack of NADH oxidase activity in *L. pseudomesenteroides* is supported by two related observations. First, the yield of mannitol from fructose with *L. pseudomesenteroides* was not improved, when the cells were grown under anaerobic conditions compared with semi-anaerobic conditions. If an NADH oxidase activity was present, it would most likely be competing with mannitol dehydrogenase for the reducing equivalents (NAD(P)H) in the cells and thus, negatively affect the fructose-to-mannitol yield. Second, the disappearance of dissolved oxygen from the growth medium was notably slower with *L. pseudomesenteroides* (no activity) than with *L. mesenteroides* (detectable activity).

In the case of *L. pseudomesenteroides*, the oxygen dissolved in the growth medium was slowly replaced by carbon dioxide produced by the cells. Hence, it is speculated that the rapid decrease in dissolved oxygen, seen with *L. mesenteroides,* is due both to formation of carbon dioxide and the presence of a significant NADH oxidase activity.

Production of Mannitol in a 2-l Batch Cultivation

When *L. fermentum* was cultured at high initial fructose and glucose concentrations, efficient bioconversion of fructose into mannitol was achieved (Fig. 3). In 11 h, 193.6 g fructose was consumed by the cells resulting in production of 175.3 g mannitol.

Hence, the volumetric mannitol productivity and mannitol yield were 7.6 g/l/h and 89.6 mol%, respectively (final VOLUME=2.11 l). A maximal volumetric productivity of 16.0 g/l/h was achieved between *t*=8 h and *t*=9 h. No residual glucose was detected, when the initial fructose was depleted, as seen in experiments with low initial sugar concentrations.

Although an increased growth temperature (40 °C) was used, the yield of mannitol from fructose was not as high as expected based on earlier comparison studies (see Fig. 1). It is speculated that the high initial sugar concentrations used in this experiment most likely altered the metabolism of the cells in an unfavourable direction. In earlier studies with growing LAB cells, the best volumetric mannitol productivity achieved is 6.4 g/l/h . Hence, we now report an improvement to this level.

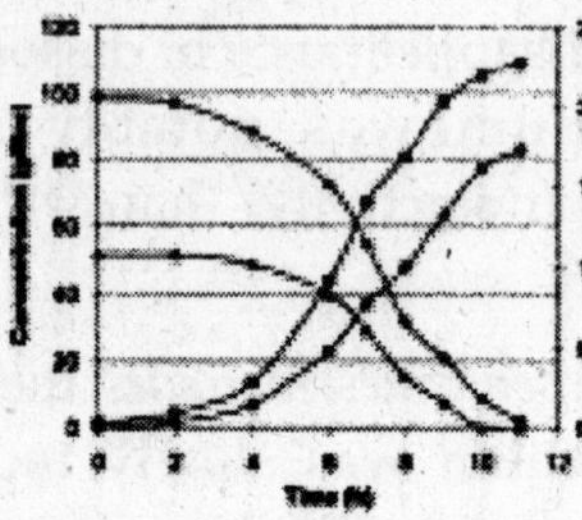

Fig. 3. Mannitol production with *Lactobacillus fermentum* NRRL-B-1932 in a 2-l batch bioreactor cultivation. Legends: *open circles,* fructose (g/l); *open triangles,* glucose (g/l); *closed circles,* mannitol (g/l); *closed rectangles,* optical density at 600 nm.

CONCLUSIONS

The ability to produce mannitol from fructose was found to vary notably among the heterofermentative LAB species studied here. In general, good mannitol productivity was favoured by high temperature and pH, whereas a good mannitol yield was favoured by low temperature and pH. In batch cultivations, the volumetric mannitol productivity is strongly influenced by the growth rate of the cells.

However, strains and species with similar growth rates are still likely to differ in mannitol production capabilities. Mannitol dehydrogenase (EC 1.1.1.67) is the key enzyme responsible for converting fructose into mannitol. Typically, among heterofermentative LAB a varying fraction of the fructose that has been transported into the cell is phosphorylated by a fructokinase enzyme to form fructose-6-P and thus, further channelled into the phosphoketolase pathway.

The 'leaking' carbon skeleton is then converted stepwise into end products such as acetic and lactic acid, ethanol, and carbon dioxide. The leakage of fructose to the phosphoketolase pathway is a serious consideration in mannitol production, mostly because fructose is relatively expensive in comparison to the final selling price of mannitol.

LAB are generally unable to produce many of the essential building blocks needed for cellular growth (amino acids, vitamins etc.) and therefore grow poorly in inexpensive mineral media. Hence, efficient growth is only achieved when the growth media is supplemented with expensive complex nutrients, such as protein hydrolysates. Taking the cost of biomass production into account, volumetric productivities typically achieved in batch and fed-batch cultivations are simply not sufficient for feasible production of mannitol. A natural way to improve the competitiveness of the bioprocess alternative would be to study the application of more advanced production methods, such as resting cells, cell-recycling, continuous cultures, and cell immobilization. Topics of this kind are currently under investigation in our laboratory. Recently, using high cell densities of *L. pseudomesenteroides* ATCC-12291 immobilized to a solid carrier, Ojamo et al. achieved a volumetric mannitol productivity of about 20 g/l/h

8

Production of Sakacin by Lactobacillus Sakei

In order to investigate factors influencing the production of the bacteriocin, sakacin P, Lactobacillus sakei CCUG 42687 was grown in a completely defined medium (DML-B) with 33 components. Although the maximum sakacin P concentration obtained was higher on a complex medium due to higher cell mass, the production per cell mass was higher in DML-B. Sakacin P was produced at 4-30 degrees C, with the highest specific production at low temperatures. More sakacin P was produced at uncontrolled pH compared with cultivation at pH 6.3. Tween-80 had a positive effect on sakacin P production, while addition of sodium chloride and trace metals had negative effects. The decrease in sakacin P concentration during the late growth and stationary phases was shown to be cell-independent and promoted at high temperature and pH. Some differences in production levels of sakacin P were found among six strains of Lactobacillus sakei tested.

INTRODUCTION

Many lactic acid bacteria produce ribosomally-synthesized antimicrobial peptides called bacteriocins, which can be divided into classes based on structure, size and mode of action (Nes et al. 1996). Class II bacteriocins comprises small, heat-stable cationic peptides which contain no modified amino acids. One subgroup, class IIa, contains bacteriocins with certain sequence motifs in their N-terminal halves and are active against the food pathogen, Listeria moncytogenes (Nes et al. 1996). For commercial use of bacteriocins in food systems, optimization of their production is necessary. Most bacteriocins are produced as primary metabolites during growth (De Vuyst et al. 1996; Nilsen et al. 1998). The lactic acid bacteria are fastidious organisms and good growth usually requires complex media. The effects of growth conditions on production have been studied for many bacteriocins in such media, i.e. amylovorin L471 (De Vuyst et al. 1996; Lejeune et al. 1998), pediocin PA-1 (Biswas et al. 1991), lactocin S (Mørtvedt-Abildgaard et al. 1995), nisin (De Vuyst & Vandamme 1993), sakacin K (Leroy & De Vuyst 1999), enterocin 1146 (Parente & Hill 1992) and enterocin A and B (Nilsen et al. 1998). Production of some bacteriocins increases under unfavourable conditions, such as at low growth rates in continuous culture and the presence of ethanol (De Vuyst et al. 1996), while production of other bacteriocins decreases under such conditions (Nilsen et al. 1998). For many bacteriocins, the concentration in the culture supernatant fluid decreases in the late stages of a batch culture. This effect has been suggested to be due to proteolytic degradation, aggregation, or adsorption to cells (De Vuyst et al. 1996).

Sakacin P is a 43 amino acid class IIa bacteriocin produced by several strains of Lactobacillus sakei (Tichaczek et al. 1992; Larsen et al. 1993; Holck et al. 1994; Trüper & De Clari 1997). The bacteriocin production is regulated by a three-component system consisting of an induction factor, a histidine kinase and a response regulator (Eijsink et al. 1996; Hühne et al. 1996; Brurberg et al. 1997).

The induction factor is a bacteriocin-like peptide secreted by the producing organism itself and serves as the signal for transcriptional activation of the genes required for bacteriocin production. Compared with other studied bacteriocins, sakacin P has higher activity against the food pathogen, L. monocytogenes, and could thus be promising for use in the food industry (Eijsink et al. 1998). Complex media do not allow examination of all individual medium factors affecting bacteriocin production. However, this is feasible using a completely defined medium, and a minimal medium for bacteriocin production can thus be established. Only a few bacteriocins from lactic acid bacteria have been studied with regard to production in a defined medium, e.g. nisin (De Vuyst 1995). This work describes the first study of bacteriocin production in a defined medium by a Lactobacillus species. Lactobacillus sakei CCUG 42687, which produces sakacin P, was studied in DML, a defined medium developed for meat lactobacilli (Møretrøet al. 1998). Various medium components and growth conditions were tested for their effect on cell growth and sakacin P production. Bacteriocin production in a defined and a complex medium was compared. Growth and production of sakacin P by five other Lact. sakei strains were also compared with Lact. sakei CCUG 42687.

MATERIALS AND METHODS

Organisms and Media

The strains used in this work are listed in Table 1. All *Lact. sakei* strains were confirmed to produce sakacin P. *Lactobacillus sakei* CCUG 42687 was the principal strain studied. The organisms were maintained as stock cultures in MRS (Oxoid) with 18% glycerol at 80 °C. Growth and sakacin P production were tested on a completely defined medium (DML, composed of 18 amino acids, three nucleotide bases, three mineral salts, seven vitamins, glucose, Tween-80 and succinate as a buffer), originally developed for six strains of meat lactobacilli including *Lact. sakei* CCUG 42687 (Møretrø*et al.* 1998). For cultivation with and without pH control, 20 and 100 mmol l^{1} succinate were used, respectively. Growth and production of sakacin P in a defined medium in a fermentor was compared with that in a complex medium (Reference medium) used for optimization of growth and sakacin P production of *Lact. sakei* CCUG 42687, with the following composition (l^{1}): 10 g yeast extract (Oxoid), 10 g tryptone (Oxoid), 30 g glucose, 0·05 g $MnSO_4.H_2O$, 0·2 g $MgSO_4.7H_2O$, 1 ml Tween-80 (Sigma) and 2·7 g KH_2PO_4 (Aasen *et al.* 2000).The effect of addition of a trace mineral solution was tested. When used, 1 ml l^{1} was added to the defined medium of a solution that contained (mg l^{1}): $FeSO_4.7H_2O$ 50, $CuSO_4.5H_2O$ 3·9, $ZnSO_4.7H_2O$ 4·4, $MnSO_4.H_2O$ 1·5, $CoCl_2.6H_2O$ 0·2 and $Na_2MoO_4.2H_2O$ 0·1. Synthetic inducing peptide (IP-673, 2 mg ml^{1} dissolved in 0·1% trifluoracetic acid) was a kind gift from Vincent G.H. Eijsink (Eijsink *et al.* 1996). Purified sakacin P (282 mg l^{1}) was obtained as described previously (Holck *et al.* 1994).

Fermentation, Microtitre Plate Cultures and Growth Measurements

To study growth and production of sakacin P in a fermentor, a CF 3000 D fermentor (CHEMAP AG, Volketswil, Switzerland) with 2 l culture volume was used. The fermentor was inoculated (1% v/v) with an overnight culture grown on DML or MRS for cultivation on defined or complex media, respectively. When the cells were grown with controlled pH, the pH was monitored and maintained by automatic addition of 2 mol l^{1} KOH for DML and 2 mol l^{1} NaOH for complex media. The fermentor was operated without aeration and with slow agitation (100 rev min^{1}) to keep the fermentation broth homogenous. The optical densities (O.D.), at 600 nm, of samples from the fermentor were measured with an Ultraspec 3000 spectrophotometer (Pharmacia Biotech Ltd, Cambrigde, UK). Cell dry mass was determined at the end of growth by washing cells once with distilled water and drying at 105 °C for 6 h.

To study the effects of the individual medium components on growth and sakacin P production, the organisms were cultured in microtitre plates with a 400 l culture volume in each well in Bioscreen C (Labsystems, Helsinki, Finland), a microtitre plate reader measuring growth as O.D. For each condition to be tested, inoculated culture was divided and transferred to 10 parallel wells, and samples of 400 l were removed from the plates during growth for bacteriocin determination. Calibration curves were made to correlate Bioscreen O.D. values, O.D. values from the Ultraspec 3000 spectrophotometer and cell dry mass (CDM). These were used to convert all the O.D. values to CDM values. The following terms are used in this work to describe the production of sakacin P: volumetric production sakacin P production per volume culture (mg l^{1}); and specific production production per cell mass (mg g^{1}).

Bacteriocin Measurement

Cell supernatant fluids were collected by centrifugation (5 min, 6000 *g*, 20 °C) and the supernatant fluids were frozen at 20 °C until bacteriocin measurement. Sakacin P was measured using a microtitre plate method described previously with *L. ivanovii* Li4(pVS2) (erythromycin- and chloramphenicol-resistant) as indicator (Axelsson *et al.* 1998). Erythromycin and chloramphenicol were added to the assay medium, both to a final concentration of 10 g ml^{1}. Thawed supernatant fluid was diluted in assay medium and assayed without filtration as the producing organism would not grow in this medium. The use of a standard of purified sakacin P of known concentration on every plate resulted in quantitative measurement (mg l 1) of sakacin P with a coefficient of variation less than 15%. The sakacin P producers most likely produce at least one additional bacteriocin (Eijsink *et al*. 1998). It is therefore important that an indicator strain is specifically sensitive to sakacin P for correct quantification. Comparisons between the activity of crude (supernatant fluid) and pure sakacin P for several strains, according to Eijsink (Eijsink, personal communication; Eijsink *et al*. 1998), showed that *L. ivanovii* Li4 was insensitive to other putative bacteriocins produced under the conditions of the assay and thus, specific for sakacin P (Axelsson and Møretrø, unpublished data).

RESULTS

Effects of Temperature and pH

Initially, NaOH was used to adjust the pH of the defined medium. However, the growth rate () and the maximum optical density at 600 nm were increased by about 10%, and the volumetric production of sakacin P was increased by 40-50%, by using KOH instead. Based on this observation, KOH was used for all pH adjustments in DML. Biotin is present in DML because of its stimulatory effect on *Lact. plantarum* and *Lact. pentosus* (Møretrø*et al*. 1998), but it could be omitted from DML without affecting growth or sakacin P production of *Lact. sakei* CCUG 42687, resulting in a minimal medium for growth and sakacin P production with 33 components (DML-B).

When *Lact. sakei* CCUG 42687 was cultured in DML-B (containing 0·3% Tween-80) in the fermentor with controlled pH 6·3, the increase in sakacin P activity paralleled the increase in cell mass, indicating primary metabolite kinetics (Fig. 1a). Growth ceased when about half the glucose had been consumed, as can be seen from the KOH consumption, which corresponded to the glucose consumption. The production of sakacin P stopped at the same time. Glucose consumption continued further until depletion, but at a lower rate. After sakacin P concentration had reached its peak, a gradual decrease in concentration was evident. This decrease continued further until the fermentation was terminated, and it is hereafter referred to as the inactivation phase (see below). In general, both volumetric and specific sakacin P production was increased by decreasing temperatures of growth (Table 2). Repeated fermentations of *Lact. sakei* CCUG 42687 in DML-B with pH 6·3 at 15 °C resulted in poor growth compared with all other conditions tested.

When *Lact. sakei* CCUG 42687 was grown in the fermentor, without pH control, in DML-B (0·3% Tween-80), growth ceased at pH 5·1-5·2. The pH continued to drop further but at a lower rate until 4·8-4·9, and glucose was not completely consumed. Sakacin P production followed the same kinetics without pH control as with controlled pH, except that there was a broader peak in total activity without pH control (Fig. 1a,b). Both volumetric and specific production of sakacin P was higher without pH control, compared with pH 6·3, at the corresponding temperatures (Table 2). The specific production at pH 6·3 was higher in DML-B than in the complex Reference medium. However, the volumetric production was higher in the complex medium due to a higher cell mass (Table 2). For culture without pH control, both volumetric and specific production were higher in DML-B compared with the complex MRS medium. Growth of *Lact. sakei* CCUG 42687 and production of sakacin P in DML-B were also tested in still cultures in tubes at 4 and 37 °C. At 4 °C, a sakacin P concentration of 2·5 mg l^{1} was obtained with a cell dry mass of 0·25 g l^{1}. No growth was obtained in DML-B at 37 °C. Growth was obtained in MRS at 37 °C but sakacin P production was negligible (less than 1 g l^{1}).

Effect of pH and Heat on the Inactivation Phase

Samples were collected, at a time near to maximum activity, from the *Lact. sakei* CCUG 42687 culture at 20 °C in the fermentor, with controlled pH 6·3 and without pH control (pH 5·1 at sampling).

The samples were subjected to various treatments as shown in Table 3. As noted above, the decrease in activity in the fermentor was fastest at pH 6·3 (Table 3). When a sample from a culture grown at controlled pH 6·3 was incubated further without pH control, the pH had decreased to 4·9 at end of the incubation and the decrease in sakacin P concentration was less than in the fermentor.

Incubation of supernatant fluid from cultures grown at pH 6·3 led to a decrease in activity similar to that in the fermentor. Heat treatment of the supernatant fluid from a culture grown at controlled pH 6·3 resulted in a 50% loss of activity after 20 min.

However, further incubation of heat-treated supernatant fluid did not lead to further loss in activity. The activity was fairly stable in all samples from a culture grown without pH control. After the inactivation phase, no sakacin P potentially adsorbed to cells could be released from cells by treatment with 100 mmol l^{1} NaCl at pH 2, according to Yang *et al.* (1992). The supernatant fluid from a culture grown at 20 °C with controlled pH 6·3 was tested for non-specific proteolytic activity using casein as a substrate, as described previously (Nrs & Nissen-Meyer 1992). No proteolytic activity was detected.

Effects of Medium Components

Tween-80 stimulates production of several bacteriocins (Parente & Hill 1992; Huot *et al*. 1996). To examine whether it has similar effects on sakacin P production, *Lact. sakei* CCUG 42687 was grown at 20 °C, in a Bioscreen C in DML-B, with Tween-80 concentrations of 0, 0·025, 0·055, 0·075, 0·11, 0·2, 0·3 and 0·5% (v/v).

No growth was obtained without Tween-80 and 0·11% Tween-80 was required for optimal growth (Fig. 2). The highest sakacin P activity was obtained with 0·3% Tween-80, and this concentration was chosen for the fermentor experiments and for further optimization in the Bioscreen C.

The effects of the concentrations of other compounds present in DML-B were tested by growing *Lact. sakei* CCUG 42687, at 20 and 30 °C, in eight different DML-B based defined media composed after a fractional factorial 2^{74} experimental design with the seven variables amino acids, vitamins, bases, salts, Tween-80, glucose and succinate. This experimental design requires the variables to be present in two concentration levels, a 'low' and a 'high' level.

The low level equals the concentrations in DML-B except 0·2% for Tween-80. The high level equals the DML-B concentration times three, except for double concentration of glucose and succinate and 0·3% for Tween-80. At 20 °C, no effects on sakacin P production were observed for any of the variables. At 30 °C, high levels of amino acids had a positive effect on both volumetric and specific sakacin P production (data not shown). The volumetric production of sakacin P at 30 °C was 40% higher in the rich medium with concentrations of all compounds at the high level, compared with DML-B (0·3% Tween-80).

The addition of sodium chloride and ethanol had negative effects on production of enterocin A and B, but for these bacteriocins, the production can be partly restored by the addition of the inducing peptide (Nilsen *et al.* 1998). To investigate whether similar effects could be obtained for sakacin P, *Lact. sakei* CCUG 42687 was grown at 20 and 30 °C in a Bioscreen C in DML-B (0·3% Tween-80) and in DML-B with added sodium chloride, ethancl and a trace mineral solution (see MATERIALS and METHODS for composition and amount used).

At 20 °C, addition of sodium chloride and ethanol reduced both growth and specific bacteriocin production, with an increased effect at higher concentrations (Table 4). Addition of trace metals had a positive effect on growth, although sakacin P production was reduced by approximately 50%. At 30 °C, the effects of sodium chloride and trace metals were essentially the same.

However, ethanol had a slightly positive effect on sakacin P production and a less negative effect on growth at this temperature. Sakacin P production was generally lower in all media at 30 °C compared with the corresponding media at 20 °C, except for the medium with 3% ethanol. No effect on any parameter was observed after adding synthetic inducing peptide (IP-673) at a final concentration of 10 ng ml^{-1} to the media containing NaCl or ethanol (data not shown).

Growth and Sakacin P Production by Other Producers

Five other sakacin P producers were tested together with *Lact. sakei* CCUG 42687 for growth and production in DML-B and MRS without pH control in a Bioscreen C. All strains grew and produced bacteriocin in DML-B and MRS at 20 °C (Table 5). *Lactobacillus sakei* VB286A grew poorly in DML-B, and specific production (defined for these five strains as sakacin P per O.D. 600 nm) was lower than in the other strains.

However, in MRS, *Lact. sakei* VB286A was the strain with the highest volumetric and specific production of bacteriocin.

For all strains, both growth and bacteriocin production were better in a defined medium where the start pH was adjusted by addition of KOH instead of NaOH, and generally, more bacteriocin was produced at 20 °C than at 30 °C (data not shown). Tween-80 was essential for growth and bacteriocin production of all strains. *Lactobacillus sakei* VB286A had a low requirement for Tween-80 compared with the other strains, and grew and produced sakacin P at a Tween-80 concentration as low as 0·01 g l^{1} (not shown).

Non-producing cells were obtained for all strains except for *Lact. sakei* VB286A, as described by Eijsink *et al.* (1996), by diluting culture grown at 30 °C 10^6-fold. Production could be restored by the addition of synthetic IP-673. There were no differences in growth between producing and non-producing cultures at 30 °C. Non-producing cells were not obtained at 20 °C, even by 10^9-fold dilution of the culture.

DISCUSSION

All the tested *Lact. sakei* strains produced sakacin P in DML-B (Table 5). Hence, there was no component present only in complex media which was essential for production. Production of bacteriocins in defined media has rarely been reported and this is the first report describing the production of a class II bacteriocin by a *Lactobacillus* species in a completely defined medium, DML-B. Many bacteriocins are likely to be produced in defined media but in order to obtain appreciable amounts, the composition of such media may require some optimization.

Using a heterologous expression system in *Lact. sakei* Lb790 for production of the class II bacteriocins pediocin PA-1, piscicolin 61, sakacin A and sakacin P (Axelsson *et al.* 1998), similar amounts of these bacteriocins were obtained in DML-B compared with MRS (Møretrø and Rosshaug, unpublished data). Thus, DML-B should be a suitable medium for the study of bacteriocin production in *Lact. sakei* and, possibly, other lactobacilli as well.

The specific production of sakacin P in DML-B was highest at low temperatures. This has previously been reported for sakacin P (bavaricin A) in complex medium (Larsen *et al.* 1993).

Other bacteriocins have also been reported to have optima for production at sub-optimal temperatures for growth (Krier *et al.* 1998; Lejeune *et al.* 1998). The low production of sakacin P at 30 °C was not increased by addition of synthetic inducing peptide. Therefore, it was unlikely that the low production of sakacin P observed at this temperature was caused by low production of the inducing peptide. De Vuyst *et al.* (1996) suggested that production of bacteriocins is increased under unfavourable growth conditions, and that low growth rates increase production.

Raising the temperature to 30 °C, and addition of trace metals, increased the growth rate but reduced the production of sakacin P, thus supporting this hypothesis. However, ethanol and sodium chloride had a negative effect on both growth rate and sakacin P production, showing that growth rate *per se* is not controlling production. It could be speculated that reduced growth rate, due to an increased energy demand (e.g. salt stress), may have a negative effect on sakacin P production, while reduced growth rate due to reduced rate of enzymatic reactions may be positive, as this could result in increased pools of essential metabolites.

The fact that amino acids appeared to be a limiting factor in DML-B for sakacin P production at 30 °C, but not at 20 °C, could support this hypothesis. At high growth rates (30 °C), the amounts of amino acids available for bacteriocin synthesis could have diminished. The low production at high temperatures could also partly be explained by increased degradation/ inactivation of the bacteriocin at high temperatures, as reported by Leroy & De Vuyst (1999) for sakacin K.

The slower growth and sakacin P production observed when NaOH was used instead of KOH for pH adjustment could indicate that the cells were stressed by sodium ions, possibly the same effect as that seen upon addition of NaCl. The medium pH adjustment by NaOH, and the use of NaOH in some stock solutions, resulted in 7-8 g l^{1} of Na^+ in the final DML medium, the same Na^+ concentration as in 2% NaCl. Aadition of inducing peptide restored some of the enterocin A and B activity in cultures where NaCl or ethanol was added (Nilsen et al. 1998).

In the present study, addition of inducing peptide could not reverse the effects on sakacin P production of NaCl or ethanol. The fact that the addition of synthetic inducing peptide had no effect on producing cultures shows that the induction of sakacin P production is an 'all or nothing' process, as suggested by Eijsink et al. (1996). This differs from the induction of enterocin A and B production, which is dependent on the dose of inducing peptide (Nilsen et al. 1998).

Tded (Nilsen et al. 1998). In the present study, addition of inducing peptide could not reverse the effects on sakacin P production of NaCl or ethanol. The fact that the addition of synthetic inducing peptide had no effect on producing cultures shows that the induction of sakacin P production is an 'all or nothing' process, as suggested by Eijsink et al. (1996). This differs from the induction of enterocin A and B production, which is dependent on the dose of inducing peptide (Nilsen et al. 1998).

The increase in sakacin P activity stopped when growth stopped, corresponding to the finding that mRNA transcripts of the sakacin P genes are present until growth stops (Brurberg et al. 1997). The fate of bacteriocin activity in the post-exponential phases varies between different bacteriocins.

For some bacteriocins, the activity decreases slightly (Parente et al. 1997) whereas for others, there is an almost complete loss of activity (Lejeune et al. 1998) or activity is stable (Nilsen et al. 1998).

The concentration of sakacin P decreased in the stationary phase, and a faster decrease was observed at controlled pH 6·3 than at uncontrolled pH (Fig. 1a,b). High pH has been shown to promote the rate of decrease in concentration for other bacteriocins as well (Kaiser & Montville 1993; Parente et al. 1997; Leroy & De Vuyst 1999).

Adsorption to cells has been shown for many bacteriocins (Yang et al. 1992; Parente & Ricciardi 1994) and has been cited as a major cause for the decrease in bacteriocin concentration in the late stages of growth and in the stationary phase (Leroy & De Vuyst 1999).

It is shown here that the decrease in sakacin P continued after the cells were removed (Table 3) and in addition, that no sakacin P activity could be recovered after the cells had been subjected to a treatment previously shown to release other bacteriocins from cells (Yang et al. 1992).

This suggests that for sakacin P, the inactivation phase is probably not due to bacteriocin adsorption to cells. The significant decrease in concentration of sakacin P during heat treatment could indicate that some chemical reactions, e.g. oxidations, may take place. Worobo et al. (1994) also reported instability of carnobacteriocins at high pH and high temperatures. No proteolytic activity on casein could be detected in the supernatant fluid, but as this is a relatively insensitive test, proteolytic degradation cannot be completely ruled out. After the initial decrease in activity, a heat-treated sample appeared to be stable.

This could indicate that proteolytic degradation is a factor in the inactivation phase. Although it has been suggested that the presence of Tween-80 prevents aggregation of bacteriocin molecules in some cases (Nissen-Meyer *et al.* 1992; Huot *et al.* 1996), aggregation of sakacin P molecules as another possible factor involved in the inactivation phase cannot be disregarded. When a sample was collected during exponential growth at pH 6·3 and the supernatant fluid was further incubated, a decrease in activity was observed (not shown).

This could indicate that the factor(s) responsible for the inactivation phase also had a negative effect during actual growth, leading to lower net production at pH 6·3 compared with growth without pH control.

It is not clear what limited the growth of *Lact. sakei* CCUG 42687 in DML-B. Only marginally higher amounts of cell dry mass were obtained using controlled pH 6·3 compared with uncontrolled pH (Table 2). In complex media with a surplus of all the medium components and controlled pH, accumulation of lactic acid will eventually stop growth. Normally, this will occur at higher concentrations of lactic acid than the concentration present in DML-B cultures when growth ceased. However, DML-B has a high osmolarity due to the high number of low molecular compounds, and it is possible that at this high osmolarity lactic acid can be inhibitory for growth at lower concentrations than in complex media.

The effects of low temperature, different Tween-80 concentrations and addition of KOH *vs* NaOH found for *Lact. sakei* CCUG 42687 were essentially all confirmed with other *Lact. sakei* strains producing sakacin P, indicating that these are general effects on the production of sakacin P in defined medium.

There was a twofold variation within the group of six strains with respect to the production level of sakacin P (Table 5). Preliminary data (A. Holck, L. Kröckel, F.K. Vogensen and K. Harmark, personal communication) suggest that all the sakacin P producers contain the same chromosomal gene cluster sakacin P (spp.) as strains Lb674 and LTH673, which have been thoroughly investigated (Hühne *et al.* 1996; Brurberg *et al.* 1997).

Variations between strains are therefore probably due to host factors other than the primary bacteriocin-related genes. Similar strain variations have been reported for other bacteriocins (Yang & Ray 1994).

CONCLUDING REMARKS

The specific production of sakacin P in the defined medium, DML-B, reached levels that were equal to or higher than production in an optimized complex medium. However, as production seemed to be correlated with growth and higher cell dry mass was obtained in complex medium, higher volumetric production was obtained in the latter type of medium.

The specific production of sakacin P was notably higher at temperatures lower than those required for optimal growth, leading also to higher volumetric production. Apart from Tween-80 and the level of amino acids (at 30 °C), no specific components in the medium were found to have a positive effect on sakacin P production. Further optimization of sakacin P production using these strains probably requires changes at the genetic level (e.g. gene dosage, strong promoters), with the aim of increasing specific production. Such studies are now being carried out in this laboratory.

9

Assessing the Suitability of Quenching Agents

Methods and Results: Tests were performed to determine the suitability of different neutralizers for a range of disinfectants, using a new method based on the Bioscreen optical density analyser. Results showed that during disinfection tests, efficacy could be over-estimated due to poor, or no, neutralization of the disinfectant after a specified time of exposure to the bacteria. The failure to distinguish adequately between bacteriostatic and bactericidal effects can lead to false results during disinfectant testing. Experiments also showed that dilution of the disinfectant, following exposure to the bacteria, was not always sufficient to stop the activity of the disinfectant for chemicals with low dilution coefficients.

Conclusions: The quench test proved to be very quick and easy to perform, with results being available within 18 h. Using the Bioscreen, the test is automated and determines whether dilution into a particular neutralizer is able to inactivate a disinfectant within 30 s.

Significance and Impact of the Study: This new approach allows the efficacy of quenching agents to be determined, prior to undertaking each disinfection study, and can help in the development of more suitable quenching solutions.

The test has also been used to find suitable neutralizers for mixtures of disinfectants which might be used during studies on synergistic biocide combinations.

INTRODUCTION

For accurate assessment of surviving organisms following contact with a disinfectant, the activity of the disinfectant must be arrested at the moment of sampling (MacKinnon 1974; Reybrouck 1979; Bloomfield 1991; Sattar et al. 1995; Kemp and Schneider 2000). In early studies on disinfection, efficacy was often over-estimated due to poor, or no, neutralization of the disinfectants after a specified time of exposure to the bacteria (Langsrud and Sundheim 1998). Russell et al. (1979) state that a failure to distinguish adequately between bacteriostatic and bactericidal effects has been responsible for inaccurate conclusions, notably with quaternary ammonium compounds. Quenching activity can be carried out by one of three methods: dilution to a sub-inhibitory level, chemical neutralization or membrane filtration (Russell 1981).

Compounds with high dilution coefficients rapidly lose their activity on dilution (Hugo and Denyer 1987). However, it is not always known whether this is sufficient to remove the disinfectant bound to viable cells. For this reason it is recommended that an appropriate neutralizing agent is incorporated, as well as dilution (MacKinnon 1974; Russell et al. 1979). An inactivating agent must possess certain properties. It must have no inhibitory effect on the microbes themselves and must very rapidly overcome the activity of the antimicrobial compound. If it combines with, and neutralizes the antimicrobial agent, the resultant substance must be non-toxic to the micro-organisms (Russell et al. 1979). The other alternative approach is membrane filtration, which is not a means of inactivation but rather, the removal of antimicrobial from the microbes. Membrane filtration normally requires several washing steps (Cremieux and Fleurette 1991) so it is not an instantaneous method for stopping the action of disinfectant against the test bacteria. It is also a rather cumbersome procedure, which is impractical for carrying out large numbers of disinfection tests.

Before starting any disinfection study, it is essential to establish the most suitable and fastest neutralizing agent. Although many authors mention the importance of the neutralization step and may list suitable chemical inactivators (MacKinnon 1974; Bloomfield 1991; Langsrud and Sundheim 1998), they rarely document the methods used to obtain this information. Russell (1981) is one of the few to do this, but the methods described only assess the quenching ability after 10-20 min in suspension, or involve the use of agar containing inactivators, which may have limited relevance to how the neutralizers would behave in solution.

Langsrud and Sundheim (1998) and Kemp and Schneider (2000) also describe methods used for testing the neutralizing capacity of their neutralizers. They added a portion of the disinfectant test solution to their neutralizer, followed by the bacterial inoculum. After 5 min, the reaction mixture was serially diluted and plated onto agar. A successful quenching agent was one that resulted in survivors (because the neutralizer would have halted the action of disinfectant prior to addition of the bacteria). This test, although effective for assessing the quenching ability of solutions, is time consuming due to the use of solid culture.

The following paper describes a new approach for testing the effect of quenching agents, based on the use of the Bioscreen Microbiological Growth Analyser. The Bioscreen is an optical density reader, which has been described previously (Lambert et al. 1998) and has been used for assessing the quality control of laboratory media (Johnston 1998) as well as for disinfectant testing (Lambert et al. 1998, 1999; Johnston et al. 2000; Lambert and Johnston 2000). This new quench test is rapid, automated by using the Bioscreen, and can determine whether dilution into a specific quenching agent is successful at fully neutralizing (within 30 s) a single compound or a mixture. These experiments also assess whether the effect of dilution alone is sufficient to inactivate a compound or a mixture.

MATERIALS AND METHODS

Preparation of Bacterial Suspensions for Quench Testing

Staphylococcus aureus ATCC 6538 was grown overnight in a shaking flask containing 80 ml Tryptone Soya Broth (TSB; Oxoid CM129) at 30°C. The culture was centrifuged (Hettich Rotina model 48R, Tuttlingen, Germany) for 10 min at 1660 *g* and the resulting cell pellet was resuspended in 10 ml 0·1% peptone water. Cell number was determined by measuring the optical density (O.D.) of cells against known standards and diluted to give 1 10^{8} ml ^{1}.

Preparation of Test Disinfectants

The following compounds were prepared in sterile distilled water at appropriate concentrations prior to use: 4-chloro- 3-cresol (CC) and phenol (Fisons); 3-cresol (*m*-c) and phenylmercuric acetate (PMA; Sigma-Aldrich); 2-phenylethanol (PEA; BDH); silver nitrate ($AgNO_3$; J.T. Baker).

Preparation of Quenching Agents

Universal quenching agent (UQA)

UQA (pH 7·0) was prepared from the following ingredients (w/v) dissolved in distilled water and autoclaved: 0·1% peptone, 0·1% sodium thiosulphate, 0·5% Tween 80 and 0·07% lecithin.

Thioglycollate Quenching Solution

A 25% (w/v) stock solution of thioglycollic acid was prepared in sterile distilled water (pH adjusted to 7·0 with NaOH). On the day of use, 100 l aliquots of this stock solution were diluted, in 8·9 ml volumes of universal quenching agent, to give a final concentration of 0·25% (w/v) thioglycollate once the disinfectant (1 ml) was added.

Testing the Efficacy of Quenching Agents

A stock solution of each disinfectant or mixture was prepared at a concentration which was anticipated to give greater than 4 log reductions in 60 min. The following were added to each of four universal containers:

(i) 10 ml of the stock disinfectant

(ii) 1 ml of the stock disinfectant into 9 ml UQA

(iii) 1 ml of the stock disinfectant into 9 ml water

(iv) 1 ml of the stock disinfectant into 9 ml thioglycollate-UQA.

Controls used were 10 ml water, 10 ml UQA and 10 ml thioglycollate-UQA (to check cell viability and from which to prepare Bioscreen calibrations). Precisely 30 s after addition of disinfectant to appropriate tubes, all of the above solutions were inoculated with 100 l of the bacterial suspension. This gave an inoculum of 10^6 ml ^{1} in each tube. After a 60 min contact time, solutions were added to TSB in a Bioscreen plate (as described by Lambert et al. 1998). Log reductions for each test sample were calculated from calibration graphs produced in the appropriate control solutions (calculations described by Lambert et al. 1998).

RESULTS

Quench Tests on Individual Compounds

Table 1 shows the log reduction results from quench tests performed on individual disinfectants. If the effect of quenching agents plus dilution was successful (i.e. inactivated the disinfectant within 30 s), there should be survival of bacteria and hence, low log reduction values in the quenched solutions (UQA or thioglycollate-UQA). The quenching process can only be said to have worked if there was inactivation (> 4 log reductions) in the undiluted disinfectant solutions. If the effect of dilution alone was sufficient to inactivate the biocide, survivors would be seen in the disinfectant solutions which were diluted in water prior to addition of bacteria.

Table 1 shows that the combination of UQA and dilution was effective at stopping the action of all biocides tested except PMA. For this agent, > 4 log reductions of the bacteria occurred even after UQA had been in contact with PMA for 30 s, suggesting that the biocide was still active. Significantly, dilution alone was sufficient to halt the action of every agent except $AgNO_3$ and PMA, which is not surprising since these two compounds have low dilution coefficients (Hugo and Denyer 1987). This means that their activity is not greatly reduced on dilution. Although the activity of $AgNO_3$ cannot be stopped by dilution alone (> 4 log reductions achieved), it is neutralized when universal quenching agent is used (0·2 log reductions).

It was found that PMA could be quenched using thioglycollate-universal quenching agent. The fact that all compounds showed > 4 log reductions when no dilution or quench was employed means that any survivors seen were attributable to the efficacy of the quenching agent or dilution.

Bacterial cells were inoculated into the three controls, 10 ml water, 10 ml universal quenching agent and 10 ml thioglycollate-universal quenching agent (to check cell viability and from which to prepare Bioscreen calibrations). Figure 1 shows that the quenching agents were not themselves toxic to the test organism.

Quench Tests on Disinfectant Mixtures

Table 2 shows that combining UQA with dilution will halt the action of all combinations except those containing PMA. Again, it is evident that mixtures containing either PMA or $AgNO_3$ cannot be inactivated by dilution alone. Thioglycollate-UQA plus dilution will inactivate all combinations used in this study without being toxic to the micro-organisms. On the basis of these results, it was decided that the combination of UQA and dilution could be used in all cases unless PMA was present in a disinfectant mixture, in which case thioglycollic acid should then be added to the quenching agent.

The Effects of using an Inappropriate Quenching Agent to Halt a Disinfection Reaction

Figure 2 shows what can happen when a quenching agent does not inactivate a compound successfully. The figure shows disinfection of *Staph. aureus* with six concentrations of PMA, which were then inadequately quenched using UQA. It appeared that after 3 min, an increase in contact time had little effect on the log reduction values. From the quench test results (Tables 1 and 2, it is known that there would have been some carry-over of active PMA into the Bioscreen plate when using universal quenching agent, causing growth inhibition in the Bioscreen. This explains why log reductions were observed, even at these low PMA concentrations.

Figure 2 shows a bacteriostatic effect from low concentrations of PMA, while Fig. 3 shows the bactericidal data revealed using an appropriate neutralizer (thioglycollate-UQA). The concentrations of PMA in Fig. 3 were 10 times greater than in Fig. 2, indicating the enhanced efficacy of the thioglycollate-UQA inactivator.

DISCUSSION

The tests performed to assess the quenching ability of different neutralizers on individual compounds showed that the combination of UQA and dilution was effective at halting the action of all biocides tested except PMA. When thioglycollate was incorporated into the quenching agent, the PMA was successfully neutralized as well. Elkhouly and Yousef (1974) also found that thioglycollate was necessary to inactivate PMA.

Langsrud and Sundheim (1998) state that universal neutralization solutions are available for inactivating disinfectants. The label 'universal' is given because such an agent will contain many neutralizing compounds, each capable of inactivating a different group of chemicals. It is unlikely that a truly universal quenching agent exists which would be capable of neutralizing all groups of disinfectants while not damaging the microbes. However, a basic formulation can be prepared for inactivating the more commonly-used groups of disinfectants, to which other neutralizing chemicals may be added when required.

Cremieux and Fleurette (1991) state that some inactivators act in a way which is purely chemical neutralization (e.g. for thiosulphate with iodic compounds or thiols with mercurials). Other neutralizers contain phospholipids such as lecithin, or non-ionic surfactants like Tween 80. Non-polar groups on these molecules often lead to the uptake of antiseptics or disinfectants (Cremieux and Fleurette 1991).

In this study, universal quenching agent, UQA, was used as the basic formulation (Lambert et al. 1998), to which other compounds were added when required e.g. thioglycollic acid for mercury compounds (Elkhouly and Yousef 1974). The UQA contains: sodium thiosulphate to neutralize chlorine, hypochlorites, iodine and aldehydes (Gross et al. 1973; Russell 1981; Hugo and Denyer 1987; Bloomfield 1991); Tween 80 for phenols, parabens and quaternary ammonium compounds, QACs (MacKinnon 1974; Hugo and Denyer 1987); lecithin which, in combination with Tween, will neutralize biguanides as well as QACs (MacKinnon 1974; Russell et al. 1979; Hugo and Denyer 1987; Bloomfield 1991).

The effect of dilution of the biocide into quenching agent may also be sufficient to halt the action of certain agents, namely phenols, cresols and alcohols (Reybrouck 1979; Russell 1981; Bloomfield 1991). This was found to be the case during this study for all compounds except those with low dilution coefficients. Sattar et al. (1995) state that they and colleagues prefer the use of dilution of the organism-disinfectant mixture at the end of the contact time rather than using a quenching agent. The present study shows that this approach will not always be effective when dealing with compounds that have low dilution coefficients.

When designing these tests, it was decided that a lower level of inoculum would be used than in a normal suspension test to minimize any effects of the microbes self-quenching the biocide (as described by Johnston et al. 2000). It was also important to establish that none of the components of the quenching system were toxic to the bacterial inoculum. For instance, Kayser and Van der Ploeg (1965) and Gross et al.

(1973) point out that sodium thiosulphate can be toxic to staphylococci. They recommend that levels of sodium thiosulphate should not exceed 0·1% (w/v). Russell (1981) states that sodium thioglycollate can be used in the inactivation of mercury compounds, but that it too can be toxic to some micro-organisms. This was not found to be the case for the strain of *Staph. aureus* used in this study when used at a concentration of 0·25%.

The consequences of using an inappropriate quenching agent were clearly demonstrated during this study, when UQA was used unsuccessfully to inactivate PMA. It was clear from the results that there was some carry-over of active PMA into the Bioscreen plate when using this universal quenching agent. This explains why inactivation of *Staph. aureus* occurred, even at these low PMA concentrations, because the residual disinfectant was acting like a preservative. Denyer (1990) states that antibacterial agents may exert both bacteriostatic and bactericidal effects, often in a concentration-dependent manner. He suggests that bacteriostatic events generally result from a form of metabolic inhibition which is released when the biocide is removed, whereas bactericidal action is caused by irreversible damage to a vital structure or function of the cell (Denyer 1995, 1998). When an effective quenching agent (containing thyoglycollate) was used against PMA, even concentrations of the disinfectant approximately 10 times higher were adequately neutralized. This illustrates the importance of identifying an appropriate neutralizer, before embarking on disinfection tests, in order to avoid false results.

The neutralizer testing approach described in this paper proved to be very quick and easy to perform, with results being available in approximately 18 h (depending on the growth rate of the test bacteria). The test is automated when using the Bioscreen and is able to establish whether dilution into a particular neutralizer can inactivate a disinfectant within 30 s. The fact that the test can be readily performed on combinations of chemicals is extremely important when seeking to screen new biocide formulations which comply with legislation and increasing environmental awareness (Rowbottom 1996). Combinations of chemicals in a synergistic system may offer some viable options (Brown and Winsley 1971; Denyer 1995) by improving efficacy, allowing the use of lower concentrations with less environmental impact, and being cheaper and safer. It is therefore vital that the appropriate neutralizers are used during these studies.

10

Persistent Listeria Monocytogenes Contamination in Food Processing Plants

Properties important in the survival of persistent and non-persistent L. monocytogenes strains in food processing plants were examined, and factors in food processing lines that predispose to persistent contamination were identified.

Persistent L. monocytogenes strains showed higher adherence levels to stainless steel surfaces than non-persistent strains after short contact times. Because, adherence to stainless steel increases resistance against sanitation procedures, efficient adherence over a short period may have an effect on the initiation of persistent plant contamination.

Differences in initial minimum inhibitory concentrations (MICs) of disinfectants observed between L. monocytogenes strains may also have an effect on the survival of these strains in food processing environments. Persistent and non-persistent L. monocytogenes strains were observed to adapt to quaternary ammonium compounds (QACs), tertiary alkylamine and sodium hypochlorite at 10°C and 37°C. Persistent and non-persistent strains adapted to similar levels. The adaptive response was observed after a 2-h sublethal exposure, indicating rapid response of the cells.

The highest increase in resistance was over 15-fold. Although the increased resistance did not exceed the concentrations of disinfectants used at food processing plants, it may influence the survival of cells when the concentration of the disinfectant is sublethal due to inadequate sanitation procedures.

Since, all disinfectants caused cross-adaptation of L. monocytogenes, maintaining a high disinfectant efficiency by rotation is difficult. The only disinfectant that L. monocytogenes was not observed cross-adapt to was potassium persulphate. However, potassium persulphate caused cross-adaptation of L. monocytogenes to the other disinfectants, which reduces the effectiveness of these agents. Cross-adaptation was not seen only to disinfectants with similar mechanisms of action but also to disinfectants with different mechanisms of action, indicating non-specific responses.

Persistent and non-persistent L. monocytogenes strains were observed in all meat and poultry processing plants. The persistent strains were often widely spread in the processing 2 plant, contaminating two or more processing lines. Important factors sustaining contamination were complex processing machines and poor compartmentalization of processing lines. The elimination of L. monocytogenes from processing machines was difficult but shown to be possible with regular and thorough disassembly and for example alkali-acid-alkali washes. Compartmentalization, especially the separation of the raw area from the post heat-treatment area, seemed to affect the contamination status of processing lines, with poor compartmentalization increasing contamination.

L. monocytogenes contamination of final food products reflects the contamination status of the manufacturing food plant. Some L. monocytogenes pulsed-field gel electrophoresis (PFGE) types were found repeatedly from the product of one producer, indicating a persistent contamination in the food plant. Some PFGE types were also found repeatedly from the products of different producers, indicating persistence of these types in several plants.

In conclusion, persistent L. monocytogenes plant contamination appears to be the result of the interaction of several different factors. Properties influencing survival, including enhanced adherence to food contact surfaces and adaptation to disinfectants, in addition to such predisposing factors in the processing line as complex processing machines and poor compartmentalization may lead to persistent L. monocytogenes plant contamination.

CONCLUSIONS

1. Persistent L. monocytogenes strains were observed to adhere to stainless steel surfaces in higher cell numbers than non-persistent strains after short contact times. Such enhanced adherence increases the possibility of survival of the persistent strains due to increased resistance against prevention methods and may have an effect on the initiation of persistent plant contamination.

   The adherence level of non-persistent strains was closer to the adherence level of persistent strains after a long contact period. It appears therefore that adherence over long contact times, in contrast to adherence over shorter periods does not explain why some strains are persistent and some non-persistent.
2. The initial resistance of persistent and non-persistent L. monocytogenes strains to disinfectants varied.

Differences in resistance to disinfectants may influence the survival of strains in food processing plants. Both persistent and non-persistent strains adapted to all disinfectants investigated, with the exception of potassium persulphate. The increase in resistance was similar for persistent and non-persistent strains and did not reach the concentrations of disinfectants used at food processing plants. However, the adaptive response may have an effect on the survival of strains when they encounter suboptimal disinfectant concentrations.

3. The persistent and non-persistent L. monocytogenes strains cross-adapted to all disinfectants except potassium persulphate. Potassium persulphate was the only agent that L. monocytogenes was observed not to adapt or cross-adapt to, and thus it seems suitable for long-term use. However, potassium persulphate did cause cross-adaptation of L. monocytogenes to the other agents. The cross-adaptive responses appeared to be nonspecific as cross-adaptation was observed between related and unrelated agents. Maintaining high disinfectant efficiency of these agents by rotation may therefore not be possible, even with agents with different mechanisms of action.

4. The PFGE types found in the raw materials had not established themselves in the post heat-treatment lines. Persistent contamination therefore appears not to be generally due to continuous recontamination by raw materials. However, the contamination may have originated from raw materials. This hypothesis is supported by several studies that show a high prevalence of L. monocytogenes in raw

materials. In addition, it is possible that had more raw material samples been investigated in Study IV persistent strains would have been found in raw materials.

Some of the strains were found only in one plant, suggesting that they may have been plant-specific. However, it is difficult to determine whether a strain is plant-specific or not. Increasing the number of samples might have resulted in some of these strains being found in other plants as well. The persistent L. monocytogenes strains were widely distributed in the food processing plants and they more often contaminated cooked products than the non-persistent strains illustrating their importance in food safety. The processing lines producing fermented products were not persistently contaminated possibly due to the presence of competing microbes on the processing surfaces.

5. Compartmentalization, and especially the separation of the raw area from the post heat- treatment area, appeared to influence the contamination status of the post heat-treatment line. Poor separation of the raw area from the post heat-treatment line seemed to increase contamination pressure. The processing machines were observed to be contaminated with and to sustain L. monocytogenes contamination. Elimination of contamination from complex processing machines was shown to be possible by thorough disassembly of the machines and targeted sanitation. However, in processing lines with inadequate compartmentalization, contamination appeared to be reintroduced at some point due to poor hygiene barriers.

6. The L. monocytogenes contamination status of food products reflects the contamination status of the corresponding food plant. Some of the foods of one producer were found to be recurrently contaminated with one PFGE type, indicating that the food processing plant or the processing line was persistently contaminated. Some of the foods produced by several producers were also recurrently contaminated with the same PFGE type, indicating that the PFGE type may have been persistent in several plants.

11

Antagonistic Activity of Staphylococcus Siderophores and Chemical Biocides

The antagonistic potential of nonpathogenic *Staphylococcus* strains against *Bacillus subtilis* wild and type strains were studied under conditions simulating a paper- and board-machine environment.

The antimicrobial activity was measured by growth inhibition in an automated turbidimeter. The antagonistic potential was compared with that of generally used chemical biocides in a paper- and board-machine environment.

The siderophore-containing extracts of *Staphylococcus* strains significantly inhibited vegetative growth of *B. subtilis* and delayed the germination of spores both in synthetic and in white-water media.

The mill strains were more resistant than type strain against *Staphylococcus* siderophores and against chemical biocides. The *Staphylococcus* siderophore-containing extracts did not interfere with the bacteriostatic effect of chemical biocides, but no synergy was detected.

The results indicate the potential for application of *Staphylococcus* siderophore-containing extracts as biocontrol agents in paper- and boardmachine environment.

INTRODUCTION

The safety of consumers, the packaging material and the environment are important considerations for food-quality packaging board and paper. The cleanliness of packaging paper and board is dependent on the quality of raw materials used, the pulping and bleaching process and hygiene at the paper and board machine. Biofilms s formed on machine surfaces may seed microbes into the product, resulting in a deterioration of its hygienic quality. Biofilm and slime in paper and board machines also cause paper breaks, holes, spots, discoloration and unpleasant odours in the product as well as microbially induced corrosion, lead-ing to significant losses in profit.

At present mainly chemical biocides are used to control microbial growth and biofilm formation on paper and board machines. Many effective applications have been developed, and treatment with chemical biocides is con-stantly being improved. The diverse microflora in mills, however, as well as environmental aspects and human safety set great demands for effective microbiocides. Bac-teria growing as a biofilm are more resistant to biocides than are planktonic bacteria, and it is possible that the use of biocides will favour biofilm bacteria over unattached organisms.

The paper machine is an open system that cannot be operated aseptically. The white water and some papermaking chemicals may contain 10^3–10^9 microorganisms per ml. The organisms usually dominant in paper products are sporeforming aerobic bacteria of the genera *Bacillus and* Paenibacillus, which are also present in liquid packaging paper and board. Bacilli and paenibacilli form heat-resistant spores, which explains their survival during the drying phase of machine operation, and results in end products with a certain microbial load. Too high a load is not desirable in the manufacture of aseptic food-packaging materials. Several members of the genus *Bacillus* also excrete potential food-degrading enzymes, eg proteases and lipolytic, cellulolytic and starch-degrading enzymes.

Since almost all aerobically growing bacteria require iron for growth, chemical complexing has been utilized in bio-cide formulations to control bacterial growth in paper- and board-machines. Synthetic chemical iron complexing agents, however, have environmentally undesirable proper-ties such as aquatic toxicity and environmental persistence. One option for biocontrol could be to use biological chelat-ing agents such as siderophores. Siderophores are defined as organic iron-specific chelators that are produced by many fungi, moulds and bacteria for ensuring their iron availability. Studies have shown that siderophores have particularly high affinity for iron when compared with inor-ganic chelates that are less specific. The mode of action can also include antibiosis, and growth inhibition can be based on the interaction of siderophores and other types of antibiosis. Staphyloferrins A and B, siderophores from *Staphylococcus hyicus,* have been isolated and charac-terized.

In addition to staphyloferrins A and *B Staphylococcus* strains produce other uncharacterized siderophores. Nonpathogenic *Staphylococcus* starter cultures are siderophore producers. *Staphylococcus* strains used as starter cultures in fermented meat products have a GRAS (Generally Regarded As Safe) status, which opens the potential for using these strains and their anti-microbial agents in biocontrol in sensitive areas such as food packaging.

The present study aims to illustrate the potential of *Sta-phylococcus* siderophores against the vegetative growth and spore germination of both *Bacillus subtilis* type and paper mill strains isolated from liquid packaging board. The experiments were performed in white-water medium and compared with the effects of generally used chemical biocides.

MATERIALS AND METHODS

Bacterial Strains

Siderophores were produced by *Staphylococcus carnosus* (VTT E-94525), *Staphylococcus* spp (VTT E-94553, VTT E-94554, VTT E-94555) strains which originated from meat and by *Staphylococcus xylosus* commercial starter (DD34 Christian Hansen, VTT E-96660). *Staphylococcus* strains were from the VTT Culture Collection and were maintained as lyophilized cells (except *S. xylosus*). *Bacillus subtilis* board mill isolates (MC-1 from liquid packaging board, AA from kraftliner) and *B. subtilis* type strain (ATCC 6051) were used as target organisms. The wild strains from the culture collection of the University of Hel-sinki, Department of Applied Chemistry and Microbiology, were stored at -196 C and maintained during the experi-ments on Trypticase Soy Agar (TSA without glucose; BBL, Cockeysville, MD, USA) slopes at 4 C.

Production and Detection of Staphylococcus Siderophores

The *Staphylococcus* strains were grown in semisynthetic minimal medium according to Raaska and Mattila-Sand-holm. All glassware was acid-washed in 6 M HCl for 24 h before use.

The minimal medium was deferrated by Chelex 100 resin to a residual content of 2.4– 2.9M Fe as determined by atomic absorption spec-trometry using the flame technique. The *Staphylococcus* inocula were grown in deferrated minimal medium for 24 h at 37 C with shaking

(150–170 rpm). For siderophore pro-duction the *Staphylococcus* strains were grown in flask cul-ture (25 ml per 100 ml) aerobically (150–170 rpm) for 2 days at 37 C after which the cells were removed by centri-fugation and culture supernatants filtered through an Ultrafee-MC filter (PLGC filter, nominal molecular weight limit 10 000; Millipore, Tokyo, Japan) to remove proteins, pooled, neutralized and stored at -25 C for subsequent assays.

Siderophores were detected in culture supernatants according to the method of Alexander and Zuberer by chrome azurol S reagent (Fluka Chemicals, Buchs, Switzerland) from samples properly diluted in MES buffer (2-[N-morpholino]ethanesulphonic acid; Sigma, St Louis, MO, USA) after 24 h incubation at room temperature using a Multiscan MCC microplate reader (Labsystems Oy, Hel-sinki, Finland) at 690 nm. The siderophore production (0 OD 690 nm) of *Staphylococcus* strains measured by chrome azurol S reagent varied 3.4–4.1/10^{10} CFU ml 1

Chemical Biocides

The commercial biocides used were Daracide ® 7849 (e.a. 10% methylenebisthiocyanate), Daracide ® 7819 (e.a. 12% 2,2-dibromo-3-nitrilopropionamide), Daracide ® 856 (e.a. 1.53% 5-chloro-2-methyl-4-isothiazolin-3-one, 0.57% 2-methyl-4-isothiazolin-3-one), Daracide ® 7848 (e.a. 1.4% 5-chloro-2-methyl-4-isothiazolin-3-one, 0.5% 2-methyl-4-isothiazolin-3-one, 7.5% glutaraldehyde), dazomet (e.a. 3,5-dimethyltetrahydro-2-thio-1,3,5-thiadiazine, Dr Ehren-storfer GmbH), 2-mercaptobenzothiazolin (Aldrich M3,30–2). Daracide ® biocides were manufactured by WR Grace Oy (Helsinki, Finland). Stock solutions of biocides (0.1%) were made in ultrapure water and filter-sterilized (0.22m; Millipore, Molsheim, France) before use. Chemical bio-cides were used at concentrations of 1 and 10 ppm of the effective substance in antimicrobial assays.

Testing of Antimicrobial Activity

White water used in testing of antimicrobial activity orig-inated from a board machine producing packaging board of bleached and unbleached kraft pulp. The white water was stored at -25 C, filtered (Whatmann 40, ash-free, Maidstone, UK), buffered and supplemented with nutrients before use. The buffers used were 0.1 M MES pH 5.5, 0.1 M PIPES (piperazine-N,N -bis-2- ethanesulphonic acid, Na2-salt; Sigma, St Louis, MO, USA) pH 7.0 and 0.1 M BIZINE (N,N-bis(2-hydroxyethyl)-glycine; Sigma) pH 8.5.

The antimicrobial activity of *Staphylococcus* culture supernatants against wild and type strains of *B. subtilis* was studied with an automated turbidometer Bioscreen (Labsystems Oy, Helsinki, Finland) [29,30]. The analyser measures microbial growth by vertical pathway and the optical changes in liquid medium are correlated with microbial counts in the samples. The optical measurements (wide-band filter) conducted at a fixed schedule throughout the run are recorded in the memory of a desk-top computer and a kinetic follow-up of the run provides a growth curve that can be analyzed by the software. The area under the growth curve was used as a measure of microbial growth, and the reduction of area (%) was used to express the growth inhibition in the presence of the antimicrobials.

B. subtilis strains were grown in Modified King's Medium B containing per litre: 20 g of protease peptone, 10 g of gly-cerol, 0.3 g of K2HPO4, 1.5 g of MgSO4, 1 g of PIPES Na2-salt and 10 g of glucose at 37 C for 24 h with shaking (120 rpm). The growth of *B. subtilis* strains in white water supplemented with 10% trypticase soy broth (TSB without glucose; BBL) or 0.3% soluble starch (May & Baker, Dag-enham, UK) was studied by dispensing 30l of a 10 1 dilution (10^6 CFU ml [1]) from overnight cultures to microti-tre plate wells with 270l of white-water media and incu-bated at 45 C for 24 h. *Bacillus* spores were produced in Modified King's Medium B supplemented with 10 mg of MnSO4·H2O L [1] at 37 C for 3 days.

The cultures were heated at 80 C for 10 min in a water bath and rapidly cooled to room temperature with cold water. Antimicrobial activity was studied by dispensing to microtitre plate wells 30l of a 10 [1] dilution of indicator strain (10^6 CFU ml 1) or spores (10^4 CFU ml [1]) with 60l of the substance to be studied and 210l of the test medium. In the control sample wells, the antimicrobial agent was replaced by an equal volume of simultaneously treated minimal medium. Effects of the antimicrobials were studied in Modified King's Medium B or in white-water medium supplemented with 0.3% starch or 10% TSB at 45 C for 24 h. All determi-nations were performed with four replicates and results were expressed as mean values.

Results

Antagonistic activity of Staphylococcus siderophores The growth of *B. subtilis* wild strain was partially (17– 46%) prevented at 45 C by the crude siderophore-contain-ing extracts of *Staphylococcus* strains tested in Modified King's Medium B (Figure 1a). The strains differed in their efficacy against *B. subtilis* wild strain; *Staphylococcus* sp strains E-94553 and E-94525 inhibited 25–46% and *S. xylosus* VTT E-96660 <20%. The pH of the test medium had little effect on the efficacy of of the *Staphylococcus* extracts, except for strain E-94553 which was more effec-tive in neutral and alkaline pH than in an acid environment.

To simulate conditions prevailing at the board machine the test was repeated in white-water medium and at a tem-perature of 45 C which is similar to that at the machine wet end. The growth of *B. subtilis* was poor in buffered (pH 7) white water, and therefore the medium was sup-plemented with TSB or soluble starch.

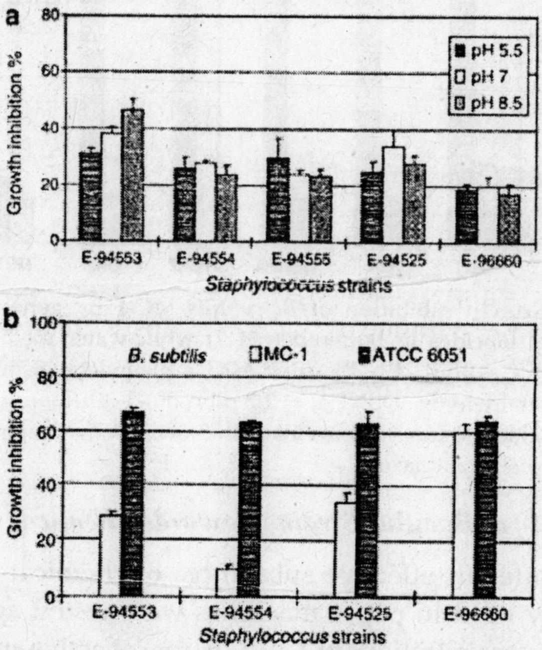

Figure 1 (a) Antagonistic activity of *Staphylococcus* crude siderophore extracts against *B. subtilis* MC-1 in Modified King's Medium B at 45 C within 24 h at different pH. Growth inhibition was expressed as the per-centage of growth area decrease in the presence of the siderophore extract compared with the control. (b) Antagonistic activity of *Staphylococcus* crude siderophore extracts against *B. subtilis* wild (MC-1) and type (ATCC 6051) strains in buffered (pH 7) white-water medium (0.3% starch) at 45 C within 24 h.

The antimicrobial activity of *Staphylococcus* siderophore-containing extracts was clear in the starch-supplemented white-water medium (Figure 1b). The growth of *B. subtilis* type strain was inhibited by all extracts 58–67%; the wild strain (MC-1) was more resistant towards three *Staphylococcus* sidero-phore extracts but similarly sensitive (60% growth inhibition) towards *S. xylosus* extract.

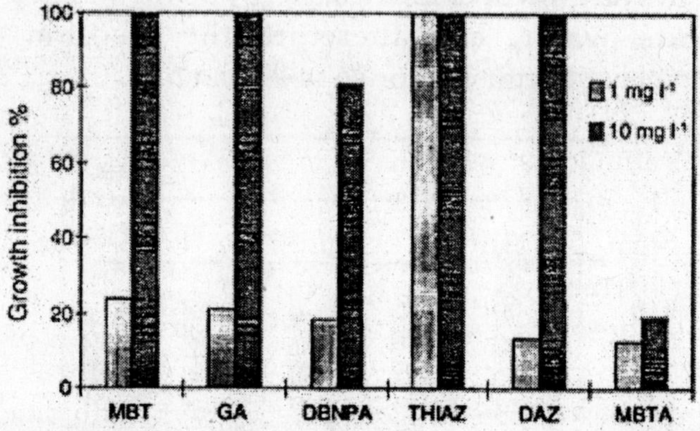

Figure 2 Growth inhibition of *B. subtilis* MC-1 by generally used com-mercial biocides in buffered (pH 7) white-water medium (0.3% starch) at 45 C within 24 h. Biocides: MBT = methylene bisthiocyanate, GA = glutaraldehyde, DBNPA = 2,2-dibromo-3-nitrilopropionamide, THIAZ = 5-chloro-2-methyl-4-isothiazolin-3-one, DAZ = dazomet, MBTA = 2-mer-captobenzothiazol

Sensitivity of Bacillus Strains Towards Chemical Biocides

Six different effective substances of chemical biocides gen-erally used in paper machines were tested against *B. subtilis* at concentrations of 1 and 10 ppm of active substance, which are concentrations used in the paper industry to con-trol microbial growth. The biocides effectively inhibited growth of *B. subtilis* MC-1 at a concentration of 10 ppm except 2-mercaptobenzothiazol (MBTA) which was less effective (Figure 2).

At 1 ppm only the combined isothiazo-lin-3-ones were effective. The sensitivities of *B. subtilis* wild and type strains to 2,2-dibromo-3-nitrilopropionamide (DBNPA) combined with *Staphylococcus* siderophores are compared in Table 1. The results show that the combined use gave no additive effect but also that the antimicrobials were not significantly antagonized by each other.

Antagonistic Activity of Staphylococcus Siderophores Against Bacillus Spores

The antagonistic activity of *Staphylococcus* siderophore-containing extracts against *B. subtilis* spores was measured and compared to that of glutaraldehyde (GA; Figure 3). The spores of the type strain were more sensitive to the sidero-phore extracts and also to GA (1 ppm) than the spores of two board mill strains (Figure 3). The germination and out-growth of spores of the type strain were inhibited 75–90%, whereas spores of the wild strains were inhibited 25–75% depending on the origin of the siderophore extract.

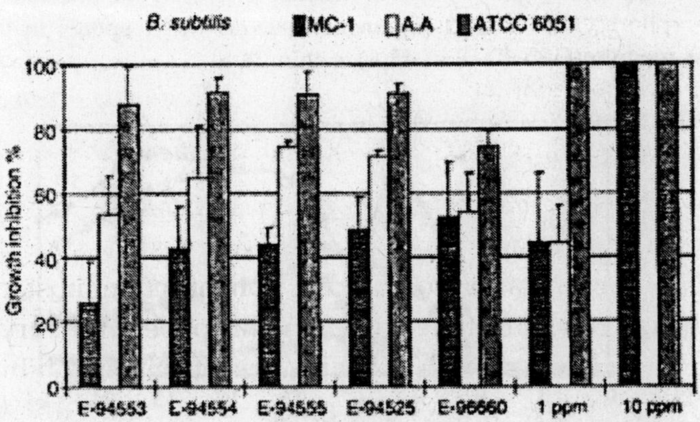

Figure 3 Inhibition of germination and early growth of *B. subtilis* spores by *Staphylococcus* siderophore crude extracts and glutaraldehyde in buff-ered (pH 7) white-water medium (0.3% starch) at 45 C within 24 h.

The *Sta-phylococcus* siderophore-containing extracts were active against *B. subtilis* wild strain spores both in acidic and alka-line environments although the activity was enhanced under acidic and neutral conditions (Figure 4). The germination of *Bacillus* spores was delayed when cultured in white-water medium supplemented with siderophore extract or GA.

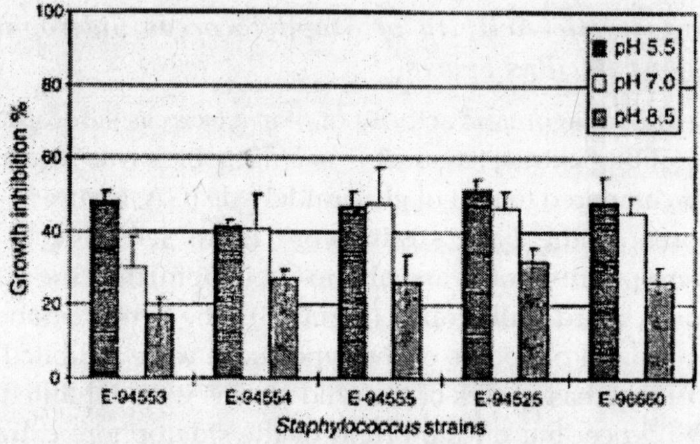

Figure 4 Effect of pH on the antagonistic activity of *Staphylococcus* sid-erophore crude extracts against *B. subtilis* MC-1 spores in white-water medium (10% TSB) at 45 C within 24 h.

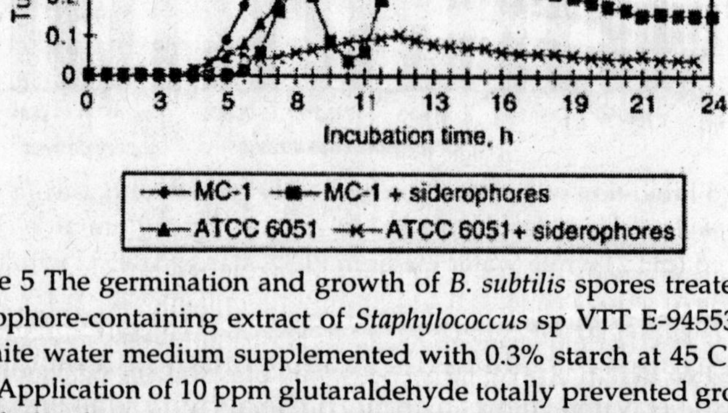

Figure 5 The germination and growth of *B. subtilis* spores treated with siderophore-containing extract of *Staphylococcus* sp VTT E-94553 strain in white water medium supplemented with 0.3% starch at 45 C within 24 h. Application of 10 ppm glutaraldehyde totally prevented growth of *B. subtilis* during the 24 h incubation. Application of 1 ppm of glutaraldehyde totally prevented the germination and growth of the type strain and delayed germination of the mill strain for 3 h.

Germination of spokes of the type strain was delayed up to 5.3 h and that of the mill strain for 1.3 h compared with the control with no siderophores. GA was very effective against the type strain, but almost as effective as sidero-phore extracts at a concentration of 1 ppm against the mill strain (Figure 5).

DISCUSSION

Increased recycling of process water by paper and board mills has decreased wastewater discharges. This may increase nutrient concentration in process water and hence increase microbial problems and the demand for biocides, contrary to what consumers and environmental authorities have been asking for.

The present paper focused on the antagonistic potential of nonpathogenic *Staphylococcus* strains against *B. subtilis* paper mill strains. The *Staphylococcus* strains used were starters from meat fermentation. *B. subtilis* was chosen as the model target organism because it is one of the important contaminants in paper and board machines and has signifi-cant heat resistance and amylolytic activity.

The effects of *Staphylococcus* siderophore crude extracts were studied at 45 C in Modified King's Medium B, which is generally used as a test medium for siderophore activity , and in white-water medium which simulates the growth environment at the board machine. The inhibitory effects of siderophores were compared with those of chemi-cal biocides generally used in paper making.

The antimicrobial effects of several of the presently used chemical biocides for slime control are believed to reside in irreversible binding of cations essential for microbial growth; eg methylenebisthiocyanate (MBT) and the thiols (eg MBTA) as well as EDTA, often used as an auxiliary biocide, are believed to deprive microbes of iron and cal-cium.

Many of the synthetic chemical biocides are not readily degraded or inactivated by the gut of warm-blooded animals or fish and therefore are inherently toxic to man and animals. Recently, methods directed at controlling microbial slimes also include the addition of microbes or enzymes in the process waters.

Many microbes excrete effective complexing agents, sid-erophores, to scavenge iron. Meat starter cultures of *Sta-phylococcus* are known to produce such siderophores and yet to be completely safe for human consumption. It is therefore assumed that *Staphylococcus* siderophores are also safe in the environment.

The present paper shows that *Staphylococcus* siderophore extracts inhibited the growth of *B. subtilis* board mill and type strains at 45 C both in complex rich medium as well as in white-water medium, the highest inhibitions being observed in low-nutritious white-water medium where the inhibition percentage of mill strain attained 60%.

The iron concentration, one of the most important controlling factors of siderophore activity in white-water medium was very low (0.02 mg L 1) com-pared with in the Modified King's Medium B (0.44 mg L 1), indicating more effective activity of *Staphylococcus* siderophores.

The culture supernatants of *Staphylo-coccus* strains were ultrafiltered before use to remove pro-teins from the samples, however, antimicrobial agents other than siderophores may have been contained in the crude extracts of the cultures used in the present study. Interest-ingly, the mill strains were generally less sensitive than the type strain both towards *Staphylococcus* siderophore-con-taining extracts and towards conventional chemical bio-cides, indicating the importance of using genuine strains isolated from the application problem area when assessing the effectivities of various antimicrobials.

Bacillus species isolated from the paper process or pro-ducts produce hydrolytic enzymes attacking raw materials of paper and board, for example starch and hemicellulose, proteins, lipids and fats. The siderophore crude extracts studied showed inhibitory activity towards amylo-lytic board mill strains of *B. subtilis,* comparable or better than some of the presently used chemical biocides in the paper industry. Staphylococcal siderophore extracts may have the potential for protection of industrial starches against colonization by amylolytic *B. subtilis.*

The crude siderophore extracts also delayed the germi-nation and outgrowth of *B. subtilis* spores. Since the spores most likely germinate and propagate during the pulping and storage phases of broke, which then seeds the machine chest and white water, siderophores may offer one option for attenuating the flow of *Bacillus* species.

The most important property of GA is its sporicidal efficacy, which is favoured with increased pH. The activities of *Staphylococcus* siderophores against the germination and outgrowth of *B. subtilis* spores were enhanced at neutral and acidic conditions.

Gram-positive microorganisms including several *Bacil-lus* species are sensitive to nisin and a range of other con-ventional bacteriocins. Furthermore, many *Bacillus* species are susceptible to EDTA and other nonspecific c chel-ators. The chelating agents remove magnesium and calcium ions from the cell membrane, disturbing the cell integrity, whereas siderophores bind iron specifically and their activity is receptor-bound. The combined use of *Staphylococcus* siderophore-containing extracts and DBNPA showed that neither interfered with the activity of the other, which would be important for the combined use of chemical biocides and antimicrobials in the wet end of the paper or board machine.

Synergism, however, has been detected between nisin and *Staphylococcus* siderophore-containing extracts. The results clearly show that *Sta-phylococcus* siderophore-containing extracts have potential for application as biocontrol agents. Hence, it would be important to further study the antimicrobial activities of sid-erophores alone or in combination with biocide activities in real paper or board machine environments and to deter-mine the economic and environmental benefits s of sidero-phore application.

12

Residual Phosphate Concentration under Nitrogen-limiting Conditions

We investigated the influence of inorganic phosphate concentration on the production of curdlan by *Agrobacterium* species. A two-step culture method was employed where cells were first cultured, followed by curdlan production under nitrogen-limiting conditions. In the curdlan production step, cells did not grow but metabolized sugar into curdlan. Shake-flask experiments showed that the optimal phosphate concentration for curdlan production was in the range of 0.1-0.3 g l^{-1}. As the cell concentration increased from 0.42 to 1.68 g l^{-1} in shake-flask cultures, curdlan production increased from 0.44 to 2.80 g 1 -1. However, the optimal phosphate concentration range was not dependent upon cell concentration. The specific production rate was about 70 mg curdlan g-cell^{-1} h^{-1} irrespective of cell concentration. When the phosphate concentration was maintained at 0.5 g 1^{-1} under nitrogen-limiting conditions, as high as 65 g 1^{-1} of curdlan was obtained in 120 h. *Journal of Industrial Microbiology& Biotechnology* (2000) 25,180-183.

INTRODUCTION

Curdlan is a water-insoluble extracellular polysaccharide com-posed exclusively of 3-1,3-linked glucose residues. It is synthesized by *Alcaligenes faecalis* var. *myxogenes and* Agro-bacterium radiobacter under nitrogen-limiting conditions. The production of curdlan has drawn considerable test because of its unique rheological and thermal gelling properties. Curdlan is used in food products such as jelly, noodles, edible fiber, and new calorie - reduced products. It is also being used to enhance the fluidity of concrete while increasing its segregational stability. In addition, curdlan might be used as a drug-delivery polymer since curdlan gel can hold and control diffusion of the drug. Furthermore, curdlan sulfate was developed as an antiviral agent able to inhibit infections by the human immunodeficiency virus. Thus, there is strong interest in reducing the manufacturing cost of curdlan.

We reported the production of curdlan by *Agrobacterium* species using sucrose as the carbon source in a two-step fed-batch technique, where cells were initially grown to substantial biomass, followed by optimizing conditions for curdlan production. A maximum titre of 64 g curdlan 1^{-1} was obtained by regulating N limitation, agitation speed and pH profile.

Phosphate concentration must also be considered because it significantly influences cell growth and product formation. Production of rhamnose - containing polysaccharide by *a Klebsiella* strain was enhanced by a reduction in the phosphate content of the medium . The production of alginic acid by *Azotobacter vinelandii* drastically increases with a decrease in phosphate concentration . In contrast, a sufficient supply of phosphate produced good yields of alginate in *Pseudomonas* strains showing growth-associated production . Other examples are a high production of polysaccharides by *a Pseudomonas* strain and maximum production of xanthan by *Xanthomonas campestris* at a high phosphate concentration.

Thus, the effect of phosphate on the production of polysaccharides is different from one strain to another.

In this study, we studied the influence of inorganic phosphate concentration on curdlan production by *Agrobacterium* species. We examined the relationship between curdlan production and residual phosphate concentration by varying cell concentrations under nitrogen-limiting conditions. We also attempted to maximize the production of curdlan in a 5-1 fermentor by using the optimal phosphate concentration.

MATERIALS AND METHODS

Microorganism and Culture Medium

Agrobacterium sp. ATCC 31750 (formerly *A. faecalis* subsp. *myxogenes)* was used. The seed culture medium (YP medium) contained 20 g sucrose, 5 g yeast extract, and 5 g bacto-peptone, pH 7.0, per liter of distilled water. The nitrogen-free medium for flask cultures contained (per liter) : 20 g sucrose, 0.5 g MgSO4·7H2O, 0.5 g K2SO4, 0.6 g CaCO3, 10 ml of a trace element solution, and varying amounts of Na2HPO4·12H20 to give different concentrations of phosphate. The composition of the trace element solution was 5 g FeSO4·7H2O, 2 g MnSO4•H2O, 1 g CoCl2•6H20, 1 g ZnClz per liter of 0.1 M HCI. Sucrose, MgSO4·7H2O, and the trace element solution were sterilized separately. Initial medium pH was adjusted to 6.0. Five-liter jar fermentation medium contained (per liter): 140 g sucrose, 4.42 g NH_4C1, 0.5 g MgSO4·7H2O, varying amounts of KH_2PO_4 and 20 ml of the trace element solution.

Flask Culture

A two-step cultuie technique was used to examine the effect of phosphate on curdlan production. Cells were first grown at 30°C for 17 h in 500 -ml baffled flasks containing 100 ml of YP medium on a rotary shaker at 150 rpm. Then, the appropriate amount of cells (42-168 mg dry weight harvested by centrifuging at 5000xg for 15 min) was transferred to 100 ml of the nitrogen-fire fermentation medium. Further cultivation was done at 30°C on a rotary shaker at 150 rpm.

Five-liter Fermentor Cultures

Curdlan production was studied in a 5-1 jar fermentor (Korea Fermentor Co., Incheon, Korea) equipped with a dissolved oxygen analyzer and a pH controller. The inoculum (300 ml grown for 17 h in shake-flask culture as described previously) was transferred into the fermentor containing 2.71 of the fermentation medium. Culture temperature was controlled at 30°C. Agitation speed was maintained at 600 rpm, and the aeration rate was 0.5 vvm. The pH was controlled at 7.0 with 3 M NaOH during the cell - growth step, and then shifted to 5.5 with 3 M HCl at the time of nitrogen limitation.

Analytical Methods

For measurement of dry cell mass, 10 ml of sample was mixed with 15 ml of 0.5 N NaOH. The supernatant was removed by centrifugation. The aliquot was washed with distilled water twice, and the dry cell mass was measured after drying it overnight at 80°C. For the analysis of curdlan, 1 ml of sample was mixed with 15 ml of 3 N NaOH solution, and incubated at room temperature for 30 min to dissolve the curdlan. After centrifuging the mixture at 5000 xg for 15 min, the curdlan in the supernatant was precipitated by adding 15 ml of 3 N HCL The precipitated curdlan was harvested by centrifuging the suspension at 5000 xg for 15 min, and it was then washed three times with distilled water to remove salts. Curdlan concentrations were determined by measuring the dry weight after drying overnight at 80°C. Sucrose concentration was measured with a modified dinitrosalicylic acid method. Ammonium concentrations were determined using the indophenol method. The phosphate concentration was determined with the ascorbic acid method.

RESULTS

Effect of Extracellular Phosphate Concentration on Curdlan Production

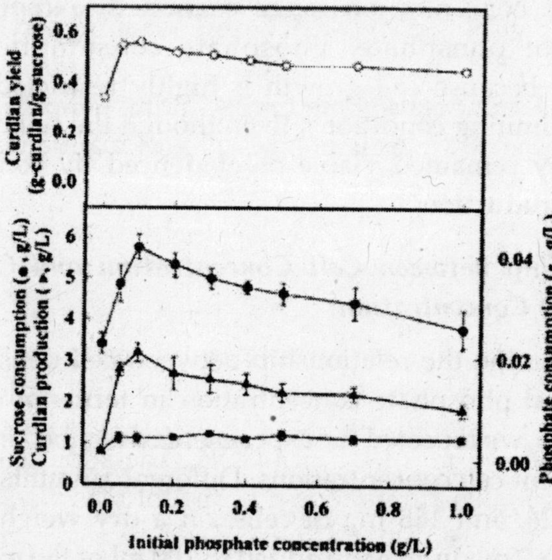

Figure] Effect of phosphate concentration on curdlan production in nitrogen-free medium. A two-step culture technique was employed. Cells were cultivated in the YP medium for 17 h at 30°C. Then, cells (84 mg dry weight) harvested by centrifugation at 5000xg for 15 min were suspended in 100 ml of the nitrogen-free medium containing different concentrations of phosphate. Further cultivation was performed for 24 h at 30°C. Experiments were done in triplicate.

A two-step culture technique was used to examine the effect of phosphate on curdlan production. Cells grown in a nutrient-rich medium (YP medium) were transferred to the nitrogen-fire fermentation medium containing various concentrations of phos-phate. As shown in Figure 1, curdlan production was highest (3.1 g 1^{-1}) at a phosphate concentration of 0.1 g 1^{-1}, increasing significantly as phosphate concentration increased from 0 to 0.1 g 1^{-1}. Curdlan production gradually decreased with further increases in phosphate concentrations.

Sucrose consumption showed the same pattern as curdlan production. Thus, the curdlan production yield from sucrose was around 0.5 g curdlan g -1 sucrose over the phosphate concentration range examined, except in the absence of phosphate. Phosphate consumption was negligible because cell growth is highly restricted under nitrogen-limiting conditions. Even though the cells did not grow, they remained viable as evidenced by continuing curdlan production.

Relationship between Cell Concentration and Optimal Phosphate Concentration

To examine the relationship between cell density and the optimal phosphate concentration in terms of curdlan production, we repeated the experiments described in Figure 1 at different cell concentrations. Different amounts of cells (42, 84, 126, and 168 mg of cells, on a dry weight basis) grown in YP medium were added to 100 ml of the nitrogen-free fermentation medium containing different phosphate levels and further cultured. The optimal phosphate concentration proved to be constant, within the range of 0.1-0.3 g 1^{-1}, which demonstrated that it was independent of cell density (Table 1). Curdlan production correspondingly increased from 0.64 to 2.86 g 1^{-1}, being directly proportional to cell concentration.

Batch Fermentation of Curdlan at Different Concentrations of Phosphate

To examine the pattern of curdlan production at different phosphate concentrations in more detail, batch fermentation was carried out in a 5-1 jar fermentor. The conditions used during the batch fermentation were different, in that the nitrogen-limited condition was allowed to develop as the cells consumed the nitrogen source, whereas in flask cultures, a nitrogen-free medium was used in the second step.

***Table l Optimal phosphate concentration range for curdlan production by* AQrrobacterium *species at different cell concentrations*[$]**

Cell concentration (g/1)	Optimal phosphate concentration° (g/1)	Curdlan production (g/1)	Specific production rate (mg curdlan/g cell/h)
0.42	0.1-0.2	0.64-0.72	63 -71
0.84	0.1-0.3	1.48-1.56	73 -77
1.26	0.1-0.3	1.85-2.26	61-75
1.68	0.1-0.3	2.80-2.86	69-70

[$]Experimental conditions are the same as described in Figure 1 except for the culture time (12 h). The predetermined amount of cells was transferred to flasks containing the nitrogen-free medium plus the different concentrations of phosphate ranging from 0.0 to 1.0 g/1.

[b]Cell concentration was determined by dry weight. One gram of cells represents 9x 10^{11} cells.

[°]Optimal phosphate concentration indicates the concentration range where curdlan production was the highest.

It is important to determine the initial concentration of the nitrogen source because it is the limiting factor for cell growth. In this regard, we examined cell growth at various ammonium concentrations using a Bioscreen analyzer system. Cell growth rate decreased as the ammonium concentration increased (Figure 2). However, since higher cell concentration produced more curdlan, as shown in Table 1, we chose an ammonium concentration of 1.6 g 1^{-1} to provide an appropriate cell concentration whilst minimizing the inhibitory effect of the ammonium ion.

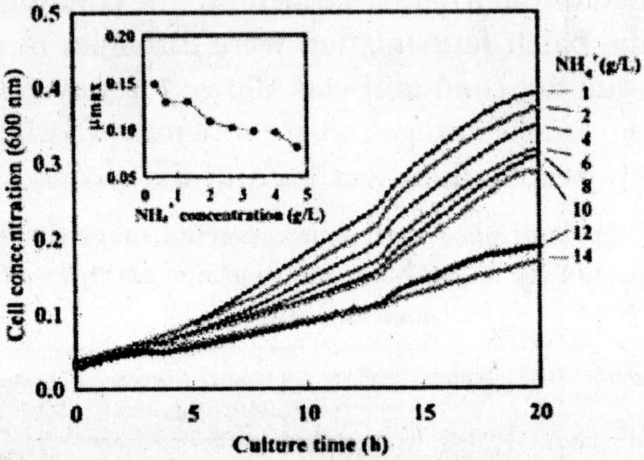

Figure 2 Effect of ammonium concentration on cell growth. Cell growth was measured by monitoring the optical density of the culture broth at 600 nm by using a Bioscreen analysing system (Labsystems, Helsinki, Finland). Cells were cultured for 20 h at 30°C. The basal medium compositions was (per liter): 20 g sucrose, 3.77 g Na_2HPO_4, 0.5 g $MgSO_4 \cdot 7H2O$, 0.5 g K_2SO_4, 10 ml of a trace element solution, and different concentrations of NH4C1. Inset in the figure indicates the variation of the maximum specific growth rate (μmax).

Once the optimum ammonium level was decided, batch fermentation was carried out at different phosphate levels. Cell and ammonium concentrations, phosphate consumption, and curdlan production levels during the batch fermentation are illustrated in Figure 3. Cell growth was proportional to ammonium consumption, and independent of phosphate concentration, which indicated that phosphate does not inhibit cell growth. When the cell concentration was at a maximum (12 g 1^{-1}), about 0.5 g 1^{-1} of phosphate was consumed, which was equivalent to a cell yield of 24 g of cells per gram of phosphate. Curdlan production began when the ammonium concentration had been significantly depleted.

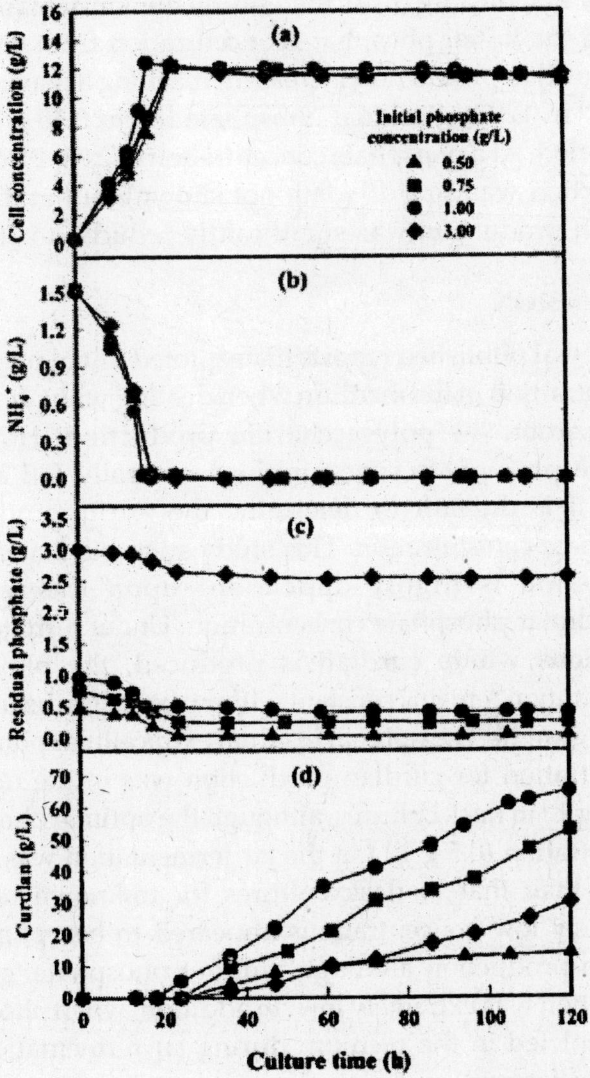

Figure 3. Batch fermentation profiles with different phosphate concentrations in a 5-1 jar fermentor. (a) Cell growth, (b) ammonium concentration, (c) phosphate consumption, and (d) curdlan production. Details are described in Materials and Methods.

When the initial phosphate concentration was 0.5 $g l^{-1}$, it decreased to negligible levels, and curdlan production was as low as 15 $g\ l^{-1}$ at the end of the fermentation. On raising the initial phosphate concentration from 0.5 to 1.0 $g\ l^{-1}$, curdlan production increased, reaching a maximum of 65 $g\ l^{-1}$ in 120 h. Residual phosphate fell to 0.50 $g\ l^{-1}$ over this period. At a phosphate concentration of 2.0 $g\ l^{-1}$, curdlan production was 54 $g\ l^{-1}$ (data not shown), but at 3.0 $g\ l^{-1}$, curdlan production was significantly reduced to 30 $g\ l^{-1}$.

DISCUSSION

Several published reports have quoted initial phosphate concentra-tion in the medium when dealing with phosphate optimization for polysaccharide production. However, because phos-phate concentrations naturally fall as cells grow, it is diiiicult to determine the intrinsic optimum phosphate concentration. This study suggests that curdlan production is highly dependent upon the residual extracellular phosphate concentration. Under nitrogen-free conditions where curdlan is produced, the phosphate concentration remains constant without being further utilized for cell growth. The optimal residual extracellular phosphate concentration for curdlan production was in the range of 0.1-0.3 $g\ l^{-1}$ in flask cultures, although the optimal phosphate concentration (0.5 $g\ l^{-1}$) in the jar fermentation was a little higher than that in flask cultures for unknown reasons. Relatively low concentrations appeared to be optimal for curdlan production although without phosphate, curdlan production was extremely low. In addition, when phosphate was depleted in the medium during jar fermentation, the medium foamed.

To obtain a high level of curdlan, residual phosphate concentration, which is dependent upon the amount of ammonium used for cell growth, needs to be defined. From the yield data (the cell growth yield from ammonium is 8.0, and from phosphate 24 g cells g - 1 substrate), it can be estimated that an initial ammonium concentration of 1.6 g 1^{-1} can yield 12 g 1^{-1} of cells, thus requiring 0.5 g 1^{-1} of phosphate for cell growth. Thus, the initial phosphate concentration should be 0.5 g 1^{-1} higher than the optimal residual phosphate concentration to support 12 g 1^{-1} of cells. With 1.0 g 1^{-1} of initial phosphate concentration that rendered 0.5 g 1^{-1} under nitrogen-limiting conditions, as high as 65 g 1^{-1} of curdlan in 120 h of batch culture was obtained. This optimal strategy can be used for the production of desired products under limiting conditions of certain essential nutrients such as nitrogen.

Bibliography

Blackburn, J.W.: *Environmental Biotechnology for Waste Treatment*, Plenum Press, New York, 1991.

David B.: *Life at a Small Scale: the Behavior of Microbes*, New York, Scientific American Library, 1996.

Ellen, Jo: *Medical Microbiology: a Short Course*, New York, Wiley-Liss, 1994.

Fausta, G.: *Biological Factors in Deterioration of Paper*, Rome, ICCROM, 1985.

Gibbons, N.E.: *Bergey's Manual of Determinative Bacteriology*, Baltimore, Williams & Wilkins Co., 1974.

Kenneth, Frank: *Ecology of Soil-bourne Plant Pathogens, Prelude to Biological Control*, Berkeley, University of California Press, 1965.

Leslie A. Lewis.: *A Novel Approach to the Evolution of Microorganisms*, New York, Vantage Press, 1987.

Michael J.: *Microbiology*, New York, McGraw-Hill, 1958.

Phillip E. Greeson: *Methods for Collection and Analysis of Aquatic Biological and Microbiological Samples*, Washington, D.C., 1989.

Rene, J.: *The World of Rene Dubos: a Collection from His Writings*, New York: H. Holt, 1990.

Steven C.: *Biotechnology, Microbes and the Environment*, San Francisco, CA: Center for Science Information, 1990.

Wood, E.J. Ferguson: *Microbiology of Oceans and Estuaries*, Amstedam, Elsevier Pub. Co., 1967.

Index

A

B

C

D

E

F

G

I

◆◆◆